About Island Press

Since 1984, the nonprofit organization Island Press has been stimulating, shaping, and communicating ideas that are essential for solving environmental problems worldwide. With more than 800 titles in print and some 40 new releases each year, we are the nation's leading publisher on environmental issues. We identify innovative thinkers and emerging trends in the environmental field. We work with world-renowned experts and authors to develop cross-disciplinary solutions to environmental challenges.

Island Press designs and executes educational campaigns in conjunction with our authors to communicate their critical messages in print, in person, and online using the latest technologies, innovative programs, and the media. Our goal is to reach targeted audiences—scientists, policymakers, environmental advocates, urban planners, the media, and concerned citizens—with information that can be used to create the framework for long-term ecological health and human well-being.

Island Press gratefully acknowledges major support of our work by The Agua Fund, The Andrew W. Mellon Foundation, Betsy & Jesse Fink Foundation, The Bobolink Foundation, The Curtis and Edith Munson Foundation, Forrest C. and Frances H. Lattner Foundation, G.O. Forward Fund of the Saint Paul Foundation, Gordon and Betty Moore Foundation, The JPB Foundation, The Kresge Foundation, The Margaret A. Cargill Foundation, New Mexico Water Initiative, a project of Hanuman Foundation, The Overbrook Foundation, The S.D. Bechtel, Jr. Foundation, The Summit Charitable Foundation, Inc., V. Kann Rasmussen Foundation, The Wallace Alexander Gerbode Foundation, and other generous supporters.

The opinions expressed in this book are those of the author(s) and do not necessarily reflect the views of our supporters.

THE SCIENCE OF OPEN SPACES

Theory and Practice for Conserving Large, Complex Systems

Charles G. Curtin

ISLANDPRESS

Washington | Covelo | London

Island Press is a trademark of The Center for Resource Economics.

Library of Congress Control Number: 2014955498

Printed on recycled, acid-free paper ♲

Manufactured in the United States of America
10 9 8 7 6 5 4 3 2 1

Keywords: Island Press, open spaces, large landscapes, thermodynamics, Malpai Borderlands, ecology, ecosystem, place-based, resilience, resilience theory, resilience practice, top-down approach, bottom-up approach, sustainability, sustainable approach, complexity, ecological policy design, collaborative conservation, resilience design, large-scale science, policy design, climate change mitigation, process design, landscape ecology, ecological restoration, community development, capacity building, adaptive capacity, sustainability studies, ecosystem studies, complex ecology, large landscape experimentation, adaptive governance

We shall not cease from exploration
And the end of all our exploring
Will be to arrive where we started
And know the place for the first time.
—T.S. Elliot

Contents

Acknowledgments

A number of people have inspired or informed this work, including Timothy H.F. Allen, James H. Brown, John Cook, Herman Karl, Steve Light, Will Hopkins, Bill McDonald, Lynn Scarlett, David Western, James Wilson, and many others. Anne Curtin, Michael Metivier, and Rowland Russell provided extensive editorial comments that greatly improved the manuscript. A special thanks to Jessica Parker and to my editors at Island Press, Barbara Dean and Erin Johnson, for their tireless efforts to improve the book. Judith McBean and the Piojo Ranch provided a quiet and secluded place to work in the closing phases of the project that was essential to the book's completion. Marica Rackstraw provided invaluable assistance in transforming my ideas into clear and compelling graphics. Above all, I wish to recognize Noreen and our two children, Conor and Rory, for their years of sacrifice on behalf of, and in support of, my career in conservation. Without their patience and understanding, this work would not have been possible.

Preface

This book stems from a 2009 graduate course I taught at MIT titled "Complexity, Ecology, and Policy Design." The positive response of the students, who were midcareer conservation professionals from around the world, to the core concepts that were a significant departure from typical approaches to conservation planning convinced me of the fundamental need for a book linking theory and practice that redrew institutional boundaries to include a wider array of perspectives and paradigms. This book was also inspired, in part, by Kai Lee's *Compass and Gyroscope,*[1] which addressed the foundations of adaptive management in a short and pithy, yet eminently readable, synthesis. My volume was originally intended as an update of Lee's 1993 book, but expanded to a broader consideration of conservation practice in large and complex systems.

In designing the 2009 course and in subsequent endeavors, I found that there was no single resource I could draw on that reached across disciplines and provided a unified perspective for students, scholars, and practitioners interested in large-scale, transboundary conservation. There are many excellent sources containing significant pieces of the puzzle, from the aforementioned book by Lee, to Gell-Mann's and Holland's works on complexity, Daniels's and Walker's book on collaborative decision making, Levin and Rosenzweig's writings on ecology, Berkes's and Folke's, and Walker's and Salt's work[2] on socioecological systems and resilience, and numerous others. But I could not find a single resource in a narrative style that provided a compelling overview of the integrated science and policy needed to sustain large ecosystems that would be accessible to broad audiences, while being solidly grounded in the underlying theory. Likewise, there was not a particular discipline that contained all the necessary conceptual underpinnings, although the obvious ones such as conservation biology, ecology, and resource management, all played their part. Equally important were areas not typically recognized by scholars and practitioners as being core to conservation, including insights from

business management, cognition, complexity theory, finance, thermodynamics, and a host of other disparate perspectives.

The volume's title is a play on that of Gretel Erlich's book *Solace of Open Spaces,* about ranching in Wyoming. The term *open spaces* contributed the unifying concept and metaphor to ground the crossdisciplinary synthesis I present. I decided I needed a new term for an approach to conservation that was distinct from others, for it was a necessary departure from previous perspectives. I found in my years of research and conserva-
tion action that the disciplinary boundaries that emanate primarily from academic scholarship simply do not map to the reality one encounters on the ground, especially at the large scales needed to address current environmental challenges where the scale of the process requires science that is a fundamental departure from previous approaches.

Large, complex issues such as climate change and sustainability are intrinsically different from many of the historical challenges society has faced and, as such, demand a new approach that matches the scale of the problem with that of the analysis and the solution. But this means re-
conceiving the role of science and policy and recognizing that sustaining science and other forms of knowledge gathering requires a synthesis of social and ecological perspectives. A science of open spaces is thus crucial not just for conservation, but in giving society a tool kit for gathering and applying information to the complex and multifaceted problems we currently face.

The core content of the book stems from a series of policy and re-
search experiments I have undertaken. From rangelands of the desert Southwest and overseas to fisheries in the Gulf of Maine, the narratives recounting these experiences provide an intellectual road map of my path of discovery. Though these case studies may seem radically different, their integration illustrates significant commonality to the underlying principles and shows a departure from conventional academic approaches in that they reflect not short-term analysis by an external researcher, but more than a decade of my firsthand experience.

Through these examples and review of the literature, the book can be distilled down to five essential points:

• There is an artificial distinction between theory and practice. In large and complex systems, an integration of approaches is essential for developing a robust tool kit for problem solving.

- The distinction between disciplines is highly subjective. Therefore, a synthesis of perspectives is essential for developing and sustaining large-scale science and policy.

- The challenges facing society today require a fundamentally different approach that re-scales science to address large and complex problems.

- A placed-based perspective is essential for building an open and transparent process that generates the trust and social capital essential for sustaining programs long enough and at scales large enough to address complex, multifaceted challenges.

- In large and complex systems, there is considerable uncertainty and a need to adapt to constantly evolving challenges. In this dynamic and uncertain context, continually refining knowledge and understanding is key for effective decision making.

CHAPTER ONE

Integrating Conservation and Complexity through the Perspective of Place

Earth so huge, and yet so bounded
pools of salt, and plots of land—
shallow skin of green and azure—
chains of mountains, grains of sand!

—*Alfred, Lord Tennyson,*
"Locksley Hall Sixty Years After"

Our Cessna banks into a tight turn above the East African savanna as Mount Kilimanjaro towers above, its summit rising through a layer of clouds. Below, a bright green expanse of wetland where hippos, elephants, and other wildlife wallow stands in stark contrast to the amber vastness of the plains of Amboseli National Park (fig. 1.1). As we enter the turn, Kenyan conservationist and ecologist David Western, handling the aircraft's controls, points out one specific, smaller patch of green coming into view, its sharply defined edges and geometric shape belying an electric fence designed to keep elephants and other large grazers out. It is an island of savanna surrounded by a sea of dust.

Western, one of the world's preeminent practitioners of large-scale conservation, describes the complex interplay unfolding below. Within the fenced enclosure are the remnants of native vegetation he saw when he first came to Amboseli as a student in the 1960s. The yellow "sea"

Figure 1.1. *The basin of Amboseli from the air, a sea of dust bisected by elephant tracks where only a few decades ago there was savanna. This is a poignant example of the impact of an imbalance between protected areas and surrounding culture—where the wildlife is forced into increasing concentrations in protected areas, where the very act of creating traditional parks that do not integrate local human needs can exacerbate conservation challenges.*

represents huge tracts of land denuded by elephant herds forced to concentrate within the park's borders for protection from poachers and conflicts with people outside the park's boundaries. Amboseli's establishment as a national park in 1974, intended as a solution to the problem of Africa's declining wildlife populations, has instead created a series of new challenges. The current predicament of landscape degradation—too many elephants in too small an area resulting in too little vegetation—can be traced to an initial lack of engagement with local Maasai peoples in conservation efforts.[1]

Establishing Amboseli as a park to protect wildlife has also meant the loss of traditional grazing grounds for the Maasai, whose presence and lifeways historically protected the elephants from poachers, and whose grazing cattle are documented to contribute to ecological richness.[2] With these pastoralists mostly removed from the park, or concentrated around established settlements and water wells, the ecosystem no longer supports the complex interactions between people and wildlife that have promoted biological diversity and sustained ecological and cultural processes for millennia.

After another pass over the plot, we fly west toward the escarpment of the Rift Valley. Upon leaving Amboseli, the landscape becomes rich and varied, as a visible interchange between people and their environment is revealed. The diversity of land use and vegetation lies in stark contrast to the relatively monotypic composition of the park. Beyond Amboseli is a tapestry of landscape features reflecting a complex interplay of ecology, economy, and culture. Our flight path from one to the other is in many respects similar to the trajectory of this book, a contrast between conventional approaches to conservation and resource management and dynamic large-landscape perspectives that promote more nuanced and resilient conservation strategies.

Traditionally, conservation offered straightforward park-based prescriptions for protecting vulnerable ecosystems, yet the reality of sustaining large, open spaces is that they are much more than the sum of their parts. The term *open spaces,* as I use it here, is intended to invoke not only the challenge of physical size but also of time, ecology, culture, and all elements therein. This is a fundamentally different approach to science that reconceptualizes both problems and solutions to generate more timely and effective means of addressing the vast conservation challenges we face today. An underlying issue I seek to address is that current approaches to science are extremely effective at meeting the demands of academia or agency-based careers and as such are structured around producing papers and professional advancement, but are less effective at addressing large and complex social and environmental problems. To make science more relevant at large scales means reconceiving its role and approach to make it more relevant to operating at large scales in messy systems where solutions do not break out cleanly along disciplinary lines. The following pages reflect two decades of experimentation not just with addressing large-scale conservation challenges, but also with changing the process itself to facilitate more effective problem solving.

Foundations in Complexity

Conventional ecology and conservation reward empirically based experimental design with robust quantitative results but largely ignore the larger social framework within which they are embedded. Conversely, traditional sociological approaches often discuss the need for empirical science without creating the institutions needed to develop and sustain such efforts or a means of evaluating the effectiveness of the work. What is missing is widespread application of a perspective that blends rigorous science with critical institutional factors. This has been characterized by social theorists Silvio Funtowicz and Jerome Ravetz as *post-normal science*: that is, extensive public engagement with the scientific process to address situations where "facts are uncertain, values in dispute, stakes high, and decisions urgent."[3] A science of open spaces links the post-normal paradigm with resilience and complexity-based perspectives, as well as the natural and social sciences, to examine how socioecological renewal and restoration stem from the emergent properties of particular land- and seascapes. At the same time, it identifies recurrent patterns of social and ecological interaction across a range of locales to find unifying strategies for successfully sustaining open spaces.

Nine thousand miles from Amboseli, Kenya, the Santa Fe Institute (SFI) sits on a hilltop above its namesake city, nestled among piñon pine and juniper. It is housed in a large, southwestern adobe-style structure that looks out across the Rio Grande Valley, the Jemez Mountains framing the western skyline. Although most academic disciplines focus on breaking down systems into pieces, SFI is a complex-systems think tank that focuses instead on synergy between biological and social systems. The Santa Fe Institute challenges the reductionist or positivist approach to science that has been a fundamental tenet of scholarship in the post–Enlightenment era in which we seek straightforward, quantitative solutions to complex, multifaceted problems. In the words of SFI cofounder and Nobel laureate Murray Gell-Mann,

> In a great many places in our society, including academia and most bureaucracies, prestige accrues principally to those who study carefully some aspect of the problem, while discussion of the big picture is relegated to cocktail parties. It is of crucial importance that we learn to supplement those essential specialized studies with what I call a crude look at the whole.[4]

SFI's perspective challenges assumptions inherent in Kenyan national park conservation strategy and in a host of global issues that cut across disciplines and scales. For a myopic focus on single, isolated variables in conventional science and policy is not enough to understand environmental change and the long-term dynamics of complex systems. As with SFI's work in physics and economics, its perspective on ecology and conservation includes the importance of novel outcomes that emerge from the *interaction* of variables. This means embracing an approach to science so that it takes a "crude look at the whole," and general measurements of real systems over precise data from models and microcosms. Looking at whole systems, with all of their social and ecological interweaving, demands working at large scales and across boundaries; in short, it demands working with open spaces.

This perspective of needing to focus on the interactions as much as the organisms was shared and complemented by the ecological studies undertaken in ecologist James H. Brown's lab at the University of New Mexico, where I did my postdoctoral work (in addition to time at SFI). Field studies looked at complex relationships among climate, livestock, and seed-eating desert rodents. The surprising results first showed that desertification (defined as an increase in shrubs and a decline in grasses) was not always caused by drought or overgrazing, but could be the result of high levels of winter rainfall, which favor deep-rooted, winter-active shrubs over the native, warm-season bunch grasses. In response to these initial changes in vegetation, the exclusion of small seed-eating rodents from experimental study plots had a far more dramatic impact on the vegetation than did 1,000-pound cows. This demonstrated that desertification-prone systems can flip into a range of configurations depending on myriad outcomes from herbivore-climate interactions.[5] These dynamics make clear the fundamental need to embrace, rather than avoid, complexity when undestanding ecological systems.

After finishing my postdoctoral work in 1998, I committed myself to finding professional scientific opportunities that embrace complexity and dynamic interactions at large scales. However, the kind of messy, tortuous science this entails is notoriously challenging to fund through conventional means. The usual time-honored approaches to science that emphasize publications and other academic achievements are largely incompatible with the integrated, transdisciplinary approaches needed to sustain large and complex systems. The traditional professional paradigm primarily supported single-disciplinary inquiry of relatively modest scales over large and integrated frameworks to problem solving. A new institutional form was needed to undertake this work integrating conservation and science at large scales, but how would it be achieved?

Potential professional opportunities arrived in three forms. The first was an offer to help develop the fledgling science program for the rancher-led Malpai Borderlands Group along the border between Arizona and New Mexico (Malpai is a variation on the Spanish *malpais*, meaning "bad-lands"). This vast basin and range landscape seemed the ideal opportunity to test-drive concepts emerging from the intersection of ecology and complexity theory. Meanwhile, funding from the Thaw Charitable Trust of Santa Fe provided support for an overall analysis of landscape change in the Borderlands through a repeat-photography study. A grant from the National Interagency Fire Center allowed us to develop landscape-level experiments of complex interactions. By selling our house, my family subsidized establishment of a new initiative called the Arid Lands Project, an institute committed to developing large-scale, place-based, post-normal research: a science of open spaces.

The three case studies (fig. 1.2) that ground this book are stories about the acquisition and application of knowledge. They illustrate new approaches to conservation, ecology, and policy through adaptive and collaborative frameworks for the management of large-scale systems. Overall, they demonstrate how such approaches allow local communities, scientists, and policy makers to work together and refine conservation

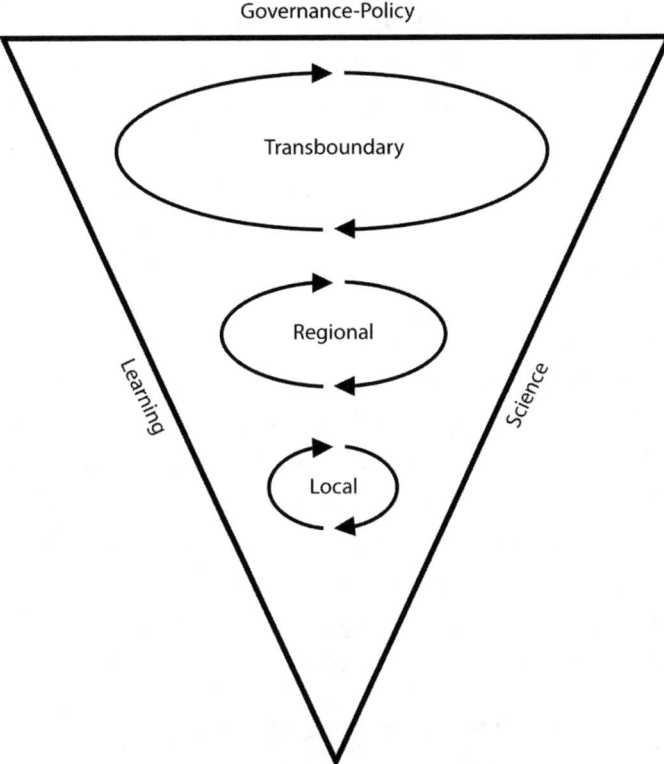

Figure 1.2. The case studies are hierarchically organized across scales to illustrate the common properties of conserving large and complex ecosystems, from the Malpai Borderlands (local), to the Gulf of Maine (regional), to a contrast between East Africa and the other two ecosystems (transboundary). Surrounding this space are three core influences on developing the capacity for adaptation: science, learning, and governance-policy. Science provides a formal means of acquiring knowledge and, perhaps most important, testing one's assumptions. Learning is essentially the way social and ecological systems evolve and adapt and is necessary for sustainability. Although governance and policy are the formal processes of applying the lessons from science and learning, these processes, in turn, are influenced by governance structures that are essentially scale-matching exercises that develop the necessary structural elements for adapting to change.

and stewardship strategies as circumstances dictate. Rather than dictating a rigid blueprint for success, these approaches leverage creativity, complexity, and social interactions to generate novel, place-based solutions.

In the Malpai borderlands, for example, ranchers and collaborators devised a different approach to the conservation of rangelands. They incorporated the region's traditional values of neighbor-to-neighbor cooperation within the broader context of modern conservation and resource management. Similarly, fishermen in the Gulf of Maine have sought to change the status quo by applying their shared understanding of the ocean to confront the paradigm of federal fishery management, thereby redefining the governance process to make it more responsive to ecological and social realities.[6] The Maasai of East Africa have, with conservation organizations such as the African Conservation Centre, built on millennia of experience to devise sustainable approaches to conserving open landscapes and the culture and ecology they encompass by devising collaborative approaches that span international boundaries, reconnecting wildlife corridors to conserve the large-scale fabric of the ecosystem. Through exploring these diverse examples, we can better understand how humans relate to their environment and witness the crucial role of place-based action in sustaining large landscapes. Each of these places exists at social and ecological boundaries, in tension zones between economic, ecological, and social forces where the stakes are high and new paradigms are essential for generating long-term solutions for sustaining open spaces.

The Perspective of Place

Fisherman Ted Ames gazed out across an immensity of uninterrupted space bounded only by the sky. This was not his home waters of Maine, or even the ocean at all. Ames's view was a vista of sand and scrub in the badlands of West Texas, mile upon mile of acacia, creosote, and mesquite rolling like waves into the distance toward Mexico and the mountains beyond. The once-great desert grasslands of this region are now covered with cactus and hummocks of sand that mark the beginnings of dune formation and an end to the ecological processes that maintained the vanished grassland landscapes of the past.

I was escorting Ames and his fellow fishermen from Maine along a 180-mile stretch of the border from El Paso International Airport to my field sites in Arizona and New Mexico, a trip that had become very familiar to me. The drive was essentially a crash course in the confluence of southwestern ecology and culture. Familiarization with the region and its biota allows visitors to better appreciate what they see when they finally arrive at the relatively intact ecosystems of the Malpai borderlands. What was remarkable about this particular trek was that of the many groups I'd presented with this montage of ecological and social change, Ted and his peers understood what they were seeing most clearly because the cycles of decline also typified their home ecosystem: the chilly waters of the western Atlantic off the coast of Maine.

The parallels between the maritime Northeast and arid Southwest continued that evening at Warner and Wendy Glenn's Malpai Ranch, outside Douglas, Arizona, as the fishermen swapped tales with the ranchers. Warner's account of roping a bear rivaled Ted's story about accidentally landing a shark that barely fit in his boat. Quirky anecdotes aside, the challenges of making a living by harvesting patchy resources across enormous open spaces were vividly clear. Though Ames's ecosystem was wet, and the Glenns' very dry, the essential cultural context and factors contributing to degrading large systems, or restoring and sustaining them, were much the same.

Ted's purpose here, so far from home, was a fishing expedition of sorts, as part of a small contingent who had come west to meet the ranchers of the Malpai Borderlands Group (MBG), learn from their experience with collaborative science and conservation, and share a bit of their own knowledge. So it was that on the following day, the New Englanders found themselves standing in front of a room full of ranchers at an MBG meeting. When former fisherman and pastor Ted Hoskins began recounting the current transformation of 400-year-old fishing traditions, it was the ranchers' turn to appreciate the challenges their guests were facing, and realize the similarities, albeit in dramatically different terrain.

The trip to MBG with the fishermen represents just one part of an odyssey involving a range of projects that spanned the desert Southwest, the marine Northeast, the Middle East, Kenya, and beyond. These experiences and partnerships form the backbone of this book, accumulated through the synthesis of the fundamental principles of developing science and policy that will sustain open spaces. Cooperation between resource users and researchers, as between the fishermen and ranchers, is emblematic of the kind of cross-cultural and cross-landscape partnerships that are necessary for effective response to environmental and social change.

As the swapping of tales between Warner Glenn and Ted Ames demonstrated, we communicate important knowledge through storytelling. Science itself is essentially an intricate series of narratives, often told through graphs and data, but essentially stories at their core. The concept of open spaces allows transboundary thinking that encourages the flow of ideas across both physical and intellectual spaces. Although this book focuses on large landscapes, the same principles apply at smaller scales and across all manner of systems. As noted by Ralph Waldo Emerson, "Life is a succession of lessons which must be lived to be understood." So too, effective science and policy require rolling up one's sleeves and investing personally in processes that are as much of an experiment as the science itself.

"Our Past Is Your Future": Comparisons of Africa and U.S. Rangelands

At Shampole Lodge in Kenya's southern Rift Valley, Malpai rancher Bill Miller stands before a group of Maasai elders, part of a delegation of ranchers and conservationists who have come to Kenya as part of an exchange of pastoralists from North America and East Africa. A big man in his sixties wearing a broad-brimmed western hat, Miller recounts the history of changes encountered by ranchers of the American Southwest. He describes the great cattle boom of the 1800s, which degraded the land and initiated a process by which large tracts were chopped into smaller holdings in a cycle of land degradation that continues to this day. Similarly, land fragmentation is the primary environmental threat faced by the Maasai, who are increasingly pushed to privatize and subdivide the land they've lived on communally for millennia.[7]

Miller's sophistication in recounting the ecological decline of his region and the roles climate, grazing, and disturbance play is impressive. "Our past is your future," he warns, recognizing the linkages between the pastoral peoples. Moses Musiaya, a Maasai elder, rises to speak and in a clear voice responds, "and our past is your future." His message is one of hope, that the Malpai ranchers can reconnect their landscape, preserve their community, and restore their ecosystems, to attain the kind of open landscapes the Maasai still maintain (fig. 1.3).

Many lessons can be drawn from this exchange. Ecological and social fragmentation are closely linked; it is nearly impossible to sustain human culture and well-being when the ecological fabric in which they are embedded is diminished. Maintaining connectivity is as important for ecosystems as it is for social systems. Finally, dialogue between people, even those from radically different cultures occupying radically different ecosystems, is essential to science and all forms of learning.

Over the last decade, natural resource users from around the globe have taken part in a series of such exchanges, dubbed *over-the-horizon learning*. Given the opportunity to live and work in unfamiliar landscapes, diverse groups have come to view their own systems through the eyes of people with different backgrounds and life experiences.[8] By examining comparisons between pastoralists, and between fishermen and pastoralists on two continents, we will see how collaborative research,

Figure 1.3. *Maasai pastoralists in East Africa have for thousands of years essentially coevolved with wildlife in the region. As such, the Maasai and their environment are a useful contrast to North America, where cattle and pastoralists are relative newcomers to the continent.*

conservation, and resource stewardship plays out across different scales. We will also see how understanding the implications of scale is key to sustaining social or ecological systems of any size. This relationship between scale and function is demonstrated especially well in the rangelands example.

Mobility and Rangelands

Rangelands are large, open ecosystems where livelihoods are gleaned from ephemeral resource patches distributed across space and time. The fundamental problem faced by pastoralists in arid and semiarid environments is maintaining sustainable access to forage that continually shifts because of uneven distribution of precipitation across the landscape. One survival strategy developed over millennia in pastoralist cultures has been to maintain mobility and cultivate reciprocal relationships among friends, neighbors, and extended families, essentially enlarging the range of options available in the face of drought or other perturbations.[9] Here, culture has been fundamentally adjusted and re-scaled to meet the challenges of ecological threats and realities, which makes both communities and ecosystems more resilient in the face of change.

However, during the last century, shifts to European land-tenure systems based on fixed land ownership (including private lands as well as public parks and reserves) have broken down these mobility-based approaches in many regions.[10] Once-great rangeland commons have been subdivided into smaller private holdings. Although subdivision of land provides stability of tenure, it also introduces socioeconomic instability: forage cannot be consistently and reliably grown on a small scale in dry climates. The scale at which rainfall influences resource patterns on the landscape is often much greater than the size of the typical subdivided parcel or landholding. Rainfall is in a general sense a proxy for forage, and forage is essential for the survival of cattle, wildlife, and associated human cultures.[11]

This disconnect between the distribution of rainfall and the scope and scale of land ownership has reduced the ecological and economic viability of many landholdings, as well as the ecological integrity of entire regions. Historic patterns of landscape fragmentation are now also compounded by rising land values, which result in the perception that the "highest and best" use of the land lies in real estate or intensive agriculture, rather than in pastoral land uses.[12] In the western United States, this has fostered the proliferation of *ranchettes* (small developed landholdings) that interrupt essential ecosystem processes such as wildfire (as discussed in the next chapter), or led to roads, fences, and other barriers that degrade and fragment an already stressed ecosystem. The goals of collaborative landowner associations such as the Malpai Borderlands Group include rebuilding social connectivity and reconnecting reciprocal approaches to land tenure. This is achieved through activities such as *grass banks,* which allow ranchers to exchange development rights for access to forage, essentially rescaling the size of their operations to make them more sustainable by allowing access to grass outside the boundaries of their land.[13] The relationship between space and socioecological function, and the integral need for building trust and social capital are strongly illustrated by the following Africa–North America rangeland comparison.

"He Who Has Been Far, Sees Far" (Maasai Proverb)

During a 2004 visit to the Mexico-U.S. borderlands, Maasai conservationist Dennis Sonkoi noted, "Many thousands of miles away from where I come from are people with similar issues and threats to their way of life." Ecologically, the rangelands of North America have undergone several major transformations, beginning with the loss of megafauna at the end of the Pleistocene more than ten thousand years ago and culminating with recent dramatic transitions in vegetation from grasslands to shrublands. East Africa, with its diverse suite of megafauna still largely intact and its longer history of human-wildlife interactions, provides a valuable context for understanding American landscapes and human-ecological relationships.[14]

The 4,000-plus-year history of domestic grazing in East Africa illustrates how human activity can contribute to healthy ecological function.[15] At night, the Maasai keep their cattle within *bomas,* temporary corrals or stockades built out of high brush and thorn to keep out predators (fig. 1.4).[16] Cattle dung accumulates there, creating nutrient-rich hotspots with gradients of human and natural disturbance radiating outward. The distribution of people and cattle across the landscape has created a patchwork of abandoned and current boma sites of varying ages that sustain

Figure 1.4. *Aerial image of Maasai lands in Kenya, illustrating both traditional and modern land use. The circle in the center is a traditional Maasai boma, where a family group lives and where cattle are held at night, concentrating dung and nutrients that have long-term positive influences on biodiversity. In the background, a rectangular field indicates that this group is becoming more sedentary. Areas of erosion are also visible, caused by the concentration of cattle around permanent settlements and water sources over an extended period of time. These shifts in land tenure are leading to landscape degradation and habitat fragmentation in a system that requires mobility and open space for its long-term survival.*

different biota in an ever-changing kaleidoscope of shifting vegetation types, all of which represent an important component of East Africa's biological diversity.[17]

In addition to maintaining environmental diversity and, therefore, integrity, mobility is key to preserving human populations. In response to dynamic climatic and ecological processes, Maasai have for many centuries used reciprocal relationships between different social groups to locate forage and move between resource patches. Even today in East Africa it is not unusual for Maasai to move their cattle hundreds of miles during times of drought as facilitated by relationships with extended family members.[18]

Similar patterns of communal land tenure and mobility appeared in the New World after the Moors' conquest of Spain in the seventh century and with the introduction of African grazing practices by the conquistadors. Beginning with Spanish colonization of the American Southwest in the 1500s and 1600s, expansive land grants were organized around large commons in which families would move their livestock in response to rainfall patterns and associated forage resources. Like the Maasai, the Spanish also used mobility to transcend variation in climate and maintain access to land and water.[19]

In the early years of widespread livestock grazing by northern Europeans in the 1800s, ranchers moved cattle vast distances to sustain herds in an open commons. Well into the twentieth century, ranching syndicates, such as Victorio Land & Cattle, spread across vast areas of the Southwest, sometimes hauling cattle to the West Coast by rail to take advantage of pasture there.[20] Likewise, the King Ranch in Texas moved cattle to well-watered Pennsylvania for access to grass and markets. But the ability to work at large scales that transcend local climate variation has largely been lost in recent decades as landscape, land tenure, and rural social systems became more fragmented.[21] A similar history is unfolding in East Africa as the Maasai are being pressured to adopt more commodity-driven approaches to livestock management and more Western forms of land tenure that heavily emphasize private property. Therefore, the outcome of fragmentation in North American open spaces provides a cautionary tale for East Africa.

The power of fragmentation to undermine culture has long been recognized. The U.S. government appears to have used fragmentation as a matter of national policy to control the Plains tribes reliant on bison. The Dawes Allotment Act of 1887 authorized the division of the land into allotments of 160 acres (too small to sustain families in most semi-arid lands) for each family and 80 acres for individuals. The Dawes Act was amended in 1891 and again in 1906 by the Burke Act. This legislation had devastating impacts on tribal lands and tribal bison herds as well as on sustainable land management in the West in general. The Dawes Act eliminated communal property holdings and led to the fencing of vast areas of open range, the loss of the tribal sovereignty and culture, and increasing reliance on the federal government.[22]

As the U.S. frontier closed in the late nineteenth century and the growing numbers of ranchers and cattle necessitated additional restrictions on land tenure, some ranchers with smaller holdings formed grazing associations. These organizations promoted the reciprocal use of resources to facilitate a more flexible response to variation in rainfall (not unlike the Spanish land grants).[23] However, in most areas, the ecological health of the rangelands still declined dramatically following the transition from subsistence agriculture to commodity production.[24] This transition was exacerbated by climate change when, in the late 1800s, the cooler, wetter climate of the Little Ice Age gave way to the warmer, drier climate of the present.[25] In the U.S. Southwest, this climatic shift led to declines in native bunch grasses that had adapted to cooler, moister conditions. Similarly, East Africa's climate is currently predicted to become warmer and drier in the coming decades, stimulating changes in vegetation and necessitating larger and more flexible approaches to land tenure.

The U.S. experience with economic change in rangelands is also significant in understanding how globalization destabilizes local communities and ecosystems. Expansion of the railroads in the post–Civil War era created access to markets and sparked a speculative boom as foreign capital flooded into the region, leading to the growth of vast cattle herds to take advantage of cheap land and plentiful grass. The cattle boom of the late 1800s led to dramatic crashes in rangeland condition and cattle population numbers in the 1890s. A second, smaller cattle boom, associated with World War I, culminated in another period of drought, as well as economic and ecological crashes in the 1920s. When many of the failed private ranches reverted to public ownership, a new system was created through the Taylor Grazing Act and other legislation of the early twentieth century wherein ranchers purchased long-term leases called *grazing allotments* on public lands managed by newly established federal land management agencies such as the Forest Service.[26]

This shift in land tenure in general, and the 1934 Taylor Grazing Act in particular, placed ranches on public lands within permanently fixed allotments, institutionalizing landscape fragmentation by dividing rangelands into arbitrary patches of pasture, each with different private management. Though grazing allotments, in theory, made ranchers more responsive to the health of a particular landscape by holding them accountable for their parcel (in the short-term, range conditions did appear to improve), the ability of individual ranchers to move herds of cattle across the landscape in response to spatial variation in rainfall was effectively ended. A key lesson for Africa is that landscape subdivision intrinsically leads to a disconnect between optimal scales for ecological sustainability and the economic constraints of private property. The two are frequently incompatible, but the need for private land to secure property rights is a current reality. To sustain open spaces, we must devise institutions that align short-term economic drivers with longer term socioecological processes. This can be done by devising land tenure institutions that provide individual or family landholdings, not through privatization, but through a system of communal shares in which lands can be sold or traded only within the community to maintain the social and ecological fabric of the system.

Human-Wildlife Interactions: Making Assets Liabilities and Liabilities Assets

Considerable progress has been made on both continents to preserve wildlife. However, African and North American rangelands continue to face similar threats such as overgrazing and landscape fragmentation, and conservation organizations on both continents struggle to change negative perceptions of human-wildlife interactions,[27] as globally wildlife continue to severely decline.[28] Conflict can be inadvertently created through policies intended to protect wildlife from people that instead transform time-honored reciprocal relationships into threat-and-response situations (fig. 1.5), often resulting in losses for both.

The crux of the problem lies in turning assets into liabilities. In Africa, governments protect wildlife by restricting hunting and establishing parks, with the unintended effect of discouraging the local people from protecting wildlife themselves. Confining wildlife to sanctuaries means that local communities receive few of the benefits from wildlife-associated tourism, and they come to view large animals, such as elephants and lions, as threats to life and property rather than important contributors to ecological and economic health. This has a reciprocal effect on ecosystems where animals are concentrated in a few parks and preserves rather than across a larger regional landscape (as in the Amboseli example at the outset of this chapter), stimulating a cascade of ecological effects that further disturb the balance between wildlife and communities. Similarly,

Figure 1.5. *In both East Africa and North America policies that make liabilities out of assets create conflict between local people and wildlife. Without benefits people will not protect their natural resources. A key facet of sustaining large landscapes is creating the incentive structure that generates the preconditions for sustainability. (Photo courtesy of Shutterfly.)*

in the United States, the federal Endangered Species Act often transforms something rare and beautiful living on the land into a direct threat to one's livelihood, rather than a source of pride.

Though both forms of regulation initially addressed very real needs to protect declining wildlife populations, a lack of social and legal integration creates dysfunctional institutions. Collaborative approaches, such as those undertaken by community-based groups—though fraught with their own challenges and complexities—are critical to the conservation of open spaces because they sustain the social fabric necessary to protect landscapes and species.[29] For example, in the Malpai borderlands described more fully in chapter 2, a habitat conservation plan encompassing numerous small parcels allowed unified conservation of endangered or threatened wildlife species within its boundaries. By sustaining diverse species, from nectar-feeding bats to montane rattlesnakes, while also contributing to the restoration of fire to maintain ecological function across the region, the habitat conservation plan allowed integration of species and landscape-level priorities within the context of the Endangered Species Act, creating a win-win situation for species, ecosystems, and people who make their living off the land. When conservation measures fairly balance human concerns for economic livelihood and the need for ecosystem-level processes such as fire that contribute to landscape health, people are more likely to support these landscape-level conservation programs.

Social Experiments: The Implications of Land Tenure for Climate Resilience

An account of the devastating 2009 drought in Kenya provides an object lesson in the costs of separating social and ecological systems as highlighted at the outset of this chapter, and why open, dynamic processes are important for maintaining open spaces. In Amboseli National Park, with its relatively monotypic vegetation and reduced forage levels caused by overpopulation of wildlife, the losses were staggering. David Western reported late in 2009:

> Nearly 15,000 animals have died of starvation since earlier this year. Wildebeest numbers fell from over 6,000 to fewer than 150, zebra from some 7,000 to 1,500 and buffalo from 600 to 185. Large numbers of elephant and many hippos have also died. Most of those losses occurred the three months between September and November and were among the biggest recorded anywhere in recent times.[30]

In contrast, the areas outside parks fared considerably better.[31] This was partly because the park failed to provide the full range of resources needed by large grazers and other animals. However, these losses are indicative of larger issues resulting from centralized conservation measures that remove humans from systems in which they are an important component. In Amboseli, the elimination of time-tested interactions between the Maasai and the landscape (as discussed above) led to a profound reduction in the ability of people, wildlife, and cattle to cope with rapid and dramatic change, a pattern graphically demonstrated by the carcasses distributed across the landscape (fig. 1.6).

Reconnecting Social and Ecological Processes

Influenced in part by the collaborative projects undertaken by ranchers in the United States, Maasai in the southern Rift Valley are working to prevent fragmentation by protecting wildlife corridors and implementing coordinated, regional approaches to science and conservation. In the African context, the close linkages between ecology, economy, and culture make for a powerful synthesis, especially given the relatively weak governance at the national level. Because wildlife is not restricted by national boundaries, Kenyan efforts are coordinated with those in Tanzania to the south.

Figure 1.6. The widespread death of large mammals during the drought in Kenya was emblematic of conservation policy failure in which there was not sufficient forage or mobility to allow animals to weather environmental variability. (Photo courtesy of Shutterfly.)

One of the most exciting recent developments in conservation has been the establishment of SORALO (South Rift Association of Land Owners), whose mission is to help Maasai reconnect their group ranches across the southern Rift Valley and adjoining landscapes.[32] The project, which began in 2004, brings together thirteen group ranches in a more than two million–acre region between Amboseli and Maasai Mara Parks into a program of "parks beyond parks."[33] The program links a number of crucial wildlife migration routes between the protected areas. The primary challenge SORALO faces lies in maintaining the connections between people and the environment that have evolved over many millennia, such as access to different habitats during the wet and dry seasons, while implementing adaptive responses to a globalized world in which land tenure is less fluid. This is especially important if communities are to move from subsistence to export economies that will provide additional financial resources to help sustain them in the face of population growth, as well as sufficient income to young people who wish to remain in the community. How can this be done without fragmenting the landhold-ings? This is a crucial question that must be addressed as we plan for conservation and rural development in the coming decades.

One of the key elements of planning for the future is developing a better understanding of the present. In Olkiramatian, southwest of Nairobi in southcentral Kenya, the South Rift Resource Centre was established through collaboration between SORALO and the African Conservation Centre. Here, the Shampole and Olkiramatian communities set aside protected areas intended to expedite community-based research. The South Rift Resource Centre is a project of the Olkiramatian Women's Group that unites communities and researchers to gather essential information for community development and resource conservation. The centre provides information on conservation and community needs while also generating income through fees paid by researchers using the site. In this way it serves the economic needs of the community while preserving ecological and cultural function.[34]

Maine Coastal Fisheries

The fishermen of coastal New England, discussed further in chapter 3, provide a parallel to pastoralists because both groups face the same fundamental challenges: matching the scale of management to the scale of the resource. Lobstermen, like ranchers and other pastoralists, repeatedly feed and handle the species they harvest. A local story a few years ago told of a Penobscot Bay lobster that a fisherman dressed up in a Barbie skirt and blouse and then released. The lobster was captured numerous times in the course of the summer, prompting much banter over the radio as different boats relayed the recapture of the fashion-conscious crustacean. This story illustrates just how many times lobster are repeatedly captured and handled prior to harvest, demonstrating how lobstering is more like the husbandry of ranching than fishing. The "bugs," as the fishermen call them, are inadvertently fed by bait, captured, and then released many times before they are finally harvested at roughly eight to ten years of age.[35]

More than a century ago, the Dawes Act and related legislation spelled the end of the open commons of the American West. Similarly, current fishery policy looks to assign fixed property rights to the world's last great commons: the oceans.[36] However, ground fisheries (harvest of bottom-dwelling fish such as cod) face the opposite challenge from that of rangelands. Landscape fragmentation and ownership patterns usually mean ranches are managed at a scale insufficient for their available resources. In contrast, the limited extent of spawning grounds and other prime fishery habitats allow large, mobile, and technologically efficient boats too much access to finite resources.[37] The result is that open marine systems are managed at a scale much larger than the distribution of resources, especially within the discrete and locally based cod and haddock populations situated on inshore spawning grounds.[38]

Examining fisheries and rangelands helps identify properties essential to sustaining the function of open spaces, despite the different specific attributes of each system. Ranchers and fishermen can learn from one another because the potential solutions for both systems lie in developing more relevant science and governance to sustain the processes that maintain social and ecological integrity. The East African application of large-scale dynamic approaches to conservation has implications for fisheries as well. For example, the Maasai have developed the concept of *olopololi* or *ol-okeri*—dynamic reserves that conserve grassland resources.[39] Olopololi recognizes the patchy, variable nature of open systems, shifting reserves in response to environmental variation. This system, of course, assumes flexible property rights that don't exist in much of the industrialized world. This concept of mobile reserves is well suited to marine systems where property rights are ascribed to resources and not real estate. Rolling closures of fishery access are somewhat analogous. Here, closures shift to track the timing of resources, for example, when fish are on spawning grounds and therefore especially vulnerable to trawling. In a system where reserve boundaries or fishing exclosures are delineated entirely by global positioning system coordinates via satellite links, there is plenty of flexibility to adaptively manage resource extraction in response to environmental variation. Olopololi can provide insights into how to track ecosystem dynamics that directly contribute to the overall function of the system. These types of innovative approaches illustrate that a community-based framework that matches the ecological scale of the resource with the social scale of tenure provides a crucial structure for conserving open spaces.

How do community-based approaches work at large scales? Studies of commons in which users communally access landscapes or resources provide important clues. In the paradox of the commons, what might lead to overexploitation and rapid decline instead often results in some of the world's most sustainable systems. Understanding why this happens is crucial to conserving open spaces.

Commons and Common Ground

In 1968, ecologist Garrett Hardin published his iconic paper "The Tragedy of the Commons," which has profoundly influenced conservation and natural resource management ever since. Hardin explained the history of the commons, pastures in medieval Britain where livestock were grazed in communal areas. In Hardin's model, which was essentially a metaphor for natural resource use in general, these areas became increasingly degraded over time as herders used more than their share to acquire forage for their livestock before their neighbor. The shared resource was quickly overexploited for individual gain. Hardin's message was that people are inherently selfish, which creates the need for a larger entity, such as the federal government, to step in and manage both resources and resource users to make sure that people do not degrade their environment.

Group and resource system boundaries and characteristics are well defined and commonly acknowledged.
The rules governing the use of collective goods are appropriate to the local context and conditions.
Collective-choice arrangements exist, with the individuals affected by the rules able to participate in their modification.
Monitoring systems are in place to generate feedback loops to promote learning by both the governing and the governed.
A system of graduated sanctions that is responsive to the seriousness of the infraction. Often locally instituted or self-enforced by the community.
Community-members have access to internal and external conflict resolution systems to address internal and external conflict between resource uses and other social actors.
External authorities recognize the rights of resource users to devise and implement their own institutions.
Governance is organized in nested-levels that recognize slow and fast ecological and social variables and integrate them into the institutional design.

Figure 1.7. *The eight core principles of sustaining common-pool resources are also relevant to sustaining large landscapes. (After Ostrom, 1990.)*

However, reality can be quite different. Some of the oldest continually used resource systems in the world, from grazing lands in East Africa to local fisheries in New England, appear to have often thrived without centralized government control. Work pioneered by Nobel Prize–winning sociologist Elinor Ostrom has shown that local engagement combined with matching the scale of the resource to the scale of the management can promote sustainability.[40]

The Maine lobster fishery demonstrates how local stewardship can generate self-organized systems in response to economic and ecological signals.[41] As with any self-sustaining system, a few basic rules govern lobstermen's access to the commons (fig. 1.7). Fixed constraints include limitations to local fishing zones and restrictions on harvest size regulated by trap numbers. This still leaves a number of strategic choices about where to fish, when to fish, and how much to fish.[42] Trap buoys serve the dual purpose of helping to capture lobster by marking the location of traps and demarcating individual fishing territories. When setting traps, a lobsterman not only assesses where the lobster are likely to be right then and in the future, but also where competitors need to be excluded to ensure that the lobsterman can follow the movement of the resource. However, although it pays to mark one's territories and intentions, it does not pay to let others know where one is actually capturing lobster. Therefore, there is also a measure of deception, or at least ambiguity, built into the selection and marking of fishing territories.

Additional trade-offs influence where gear is placed. For example, setting traps close to those of *highliners*—those lobstermen known to be top producers—may increase the catch, but setting traps too close can cause lines to become entangled, resulting in the costly loss of both time and equipment. Accessible territories, such as those close to a harbor, may be easier to work, but the propellers of recreational powerboats may inadvertently sever trap lines. Threats from other lobstermen sometimes also occur, such as when trap lines are knotted or cut as a warning for encroaching on someone's territory. Yet trap cutting is expensive for both sides because it risks retaliation and financial disaster; traps and gear cost hundreds to thousands of dollars per location. Fishermen must therefore decide between fishing the best territories, where chances of entanglements or loss of gear are high, or fishing territories with longer travel times, higher fuel costs, and lower trap success, but with less chance of conflict or loss. Complex dynamics exist in interactions between lobstermen and lobster, lobstermen and each other, and lobstermen and economic fluctuations in cost (e.g., bait or fuel) and/or income (e.g., wholesale price per pound). These interactions constitute a self-organized system in which price, catch effort, and cost combine with social interactions to sustain the fishery. For example, most lobstering communities on or close to the mainland rely on catching lobster during the summer, when the animals follow the warming water inshore to molt and breed. By contrast, Monhegan Island, located twelve miles off the coast of Maine, has a lobstering season in the winter because the lobsters reside in the island's deep and relatively warm waters during that season. This is beneficial for Monhegan lobstermen because prices are higher in the winter and the hard-shelled lobster are easier to ship and thus of greater commercial value.

Collective action is more likely to occur in an environment when it is consistent with the self-interest of the parties involved. If both the social and ecological components of open spaces such as the Gulf of Maine

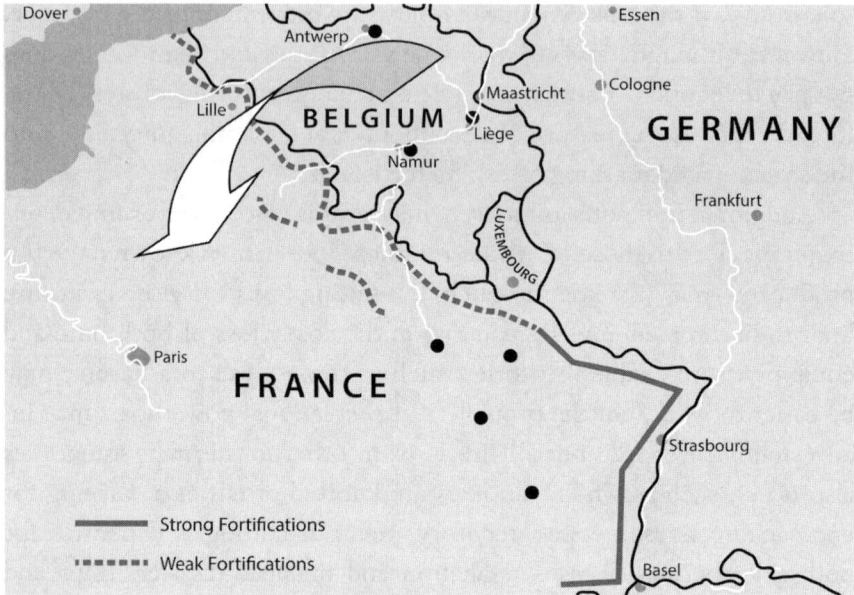

Figure 1.8. *Military strategy often fights the last war, rather than anticipating the next one. The image of the Maginot Line (the line in the center of the image) and the arrow showing the route of the German army across the Low Countries in World War II is a metaphor for understanding recurrent pathologies in conventional approaches to problem solving. The Maginot Line's entrenched fortifications, designed to avert another world war, were all but useless against the lightning armored advance of World War II. Current approaches to conservation that have emerged in response to historic demands are often not effective at addressing today's large and complex environmental and social challenges. The point of a science of open spaces is to reframe science and policy to make them more relevant in an era of increasingly rapid environmental and social change.*

(and, by extension, the borderlands and East Africa) are to be sustainable, they require self-organized local governance that is founded on scientific knowledge, respects self-interest, and encourages the continued individual and collective learning necessary to adapt to environmental and social change. The examples above provide a taste of what is possible when collaborative place-based approaches are undertaken, but they also illustrate the consequences when they are not. This book is devoted to understanding how coupling current theory and practice from disciplines ranging from organizational theory to resilience is being harnessed in new ways to promote ecologically and economically viable science, conservation, and resource stewardship (fig. 1.8).

Concluding Remarks

The preceding pages have laid the foundations for considering the interaction between ecology and society and the role of science and policy in large and complex systems. We viewed the interchange of social and ecological processes in a diversity of contexts across three scales from local to regional and international comparisons. In the next two chapters we will examine the elements of success and failure through a more detailed look at two of the case studies. These examples from rangelands and fisheries explore how the development of sustainable institutions that promote ecosystem heath and resilience occurs, and the promise and pitfalls of undertaking conservation science and resource management at the large scales essential for maintaining open spaces. These examples provide a context for viewing the challenges of sustaining large systems through theory and practice that comprise the balance of the book.

CHAPTER TWO

Experiments in Post-Normal Science in Southwestern Rangelands

I knew the wild riders and the vacant land were about to
vanish forever . . . and the more I considered the subject, the
bigger the forever loomed. Without knowing how to do it, I
began to record some facts around me, and the more I looked
the more the panorama unfolded.

—Frederic Remington

On a clear day, standing atop a windswept ridge on the southern border of the United States, one can see a hundred miles into Mexico (fig. 2.1). From the southern horizon, the Sierra Madre range extends north to meet the Rockies along the continent's spine. To the west, the low, blistering Sonoran Desert stretches from the heart of western Mexico into Arizona. To the east and southeast sits the higher, grassier Chihuahuan Desert. The Great Plains lie to the northeast, the Great Basin to the northwest. Here in the heart of a region often misconceived as singular and undifferentiated, six distinct biomes meet, as have an equally rich diversity of human communities and cultures. Home to Apache and Anasazi, to Irish and Scots, Mormons, Mennonites, Mexicans, Texans, and an increasing abundance of "snowbirds" from the north, the million-plus-acre Malpai borderlands is a human and biotic crossroads situated on a low spot on the spine of the continent. Its basin and range topography, isolated mountain ranges sometimes called *sky islands,* arid grasslands, and shrubby deserts comprise parts of two nations (Mexico and the United

Figure 2.1. The borderlands are a matrix of desert, grasslands, and montane habitats. Here, the core conservation challenge is to preserve mid-elevation desert grassland—a globally rare ecosystem that is disproportionately affected by humans and provides corridors for wildlife to move between montane habitats. Mid-elevation desert grasslands are essential to everything from montane rattlesnakes and large carnivores such as bears to iconic species such as jaguar. The view here from the southern edge of the Rocky Mountains south into Mexico shows the Sierra Madre on the horizon and the Chihuahuan Desert in the foreground.

States), four states (Chihuahua, Sonora, Arizona, and New Mexico), and numerous local, state, and federal jurisdictions. As one might expect, the challenges of sustaining conservation and land management across all of these ecological, social, and geopolitical boundaries are staggering.

Historically, these challenges have all revolved, to some degree, around the impacts of cattle ranching on these ecosystems. Wide open spaces lend themselves quite naturally to ranching-based economies. But the same ecological and climatic characteristics that have made such activities viable, including the sheer unpredictability of rainfall and the continual specter of drought, which exclude sustainable farming, also lead to concerns about degradation of natural communities. These concerns include, but are not limited to, desertification, threats to endangered species, and the prudent allocation of resources such as water. Ironically, ranching has become not only a way of life for people, through which they became intimately bound to the land, but also a lightning rod for others who view the activity as fundamentally damaging.

Since the rise of the contemporary conservation movement in the last half century, challenges to traditional ranching practices from scientists and conservationists have grown in number, scope, and insistency. Both sides often came to view the other as a threat to something they hold dear (for ranchers their land and livelihood and for environmentalists species protection and land health), and both subsequently became entrenched in their opposition to often reasonable and necessary measures. In the context of this conflict the Mexico-U.S. borderlands are a microcosm of these larger challenges that range across the West and the globe. The work of the region's rancher-led Malpai Borderlands Group is presented here because it is emblematic of both the opportunities and the complexities of conserving open spaces.

Foundations of Consensus

By the late 1980s, local borderlands rancher, poet, and Anheuser-Busch heir Drummond Hadley, used to traveling widely in both literary and ranching circles, was becoming increasingly concerned by the growing rift between his fellow cattlemen and environmentalists. Rallying cries such as "Cattle free by '93!" and "Remove the sacred cows from the public trough!" were frequently heard at the time, part of protests that sought to break down both federal and state grazing systems, which had been in place for the better part of a century, as well as the ranching communities and culture that depended on them. Debate over federal land use in the West, simmering for decades, was at the boiling point as public perceptions of grazing on public lands grew more negative.

Hadley understood environmentalists' concerns about both the scale and the effects of cattle grazing on rangelands, but he also understood firsthand the ranchers' antagonism toward conservationists and other outsiders as a defense against threats to their way of life. Amid this conflict, he also recognized that the ranching community needed to take the lead in finding solutions. So Hadley, along with Bill McDonald, Warner and Wendy Glenn, and other ranchers in the Arizona–New Mexico borderlands, turned conventional wisdom on its head by reaching out to the very same groups widely considered to be their natural enemies to form alliances in what has since been termed the *radical center*.[1] In doing so they bucked the polarized rhetoric of grazing associations versus environmental groups by bringing historically opposed groups together to find common ground.

However, Hadley's vision extended beyond bringing people together to talk about the plight of ranching. In 1993, with the help of his family, Hadley and his son Seth formed a foundation to acquire the 502-square-mile Gray Ranch from The Nature Conservancy (TNC), the nation's largest conservation organization, which itself had purchased the land in 1990. The Animas Foundation (*animas* roughly translates to "spirit" in Spanish) was named for the local community and wild and remote mountain range that is the centerpiece of the ranch. Located in extreme southwestern New Mexico, the vast property is home to an astounding diversity of plant, mammal, reptile, and bird species and is arguably the most diverse spot in North America with a long history of scientific research and natural history extending back to the nineteenth century.[2] Under TNC ownership, the Gray Ranch was the flagship of its Last Great Places campaign to preserve vast, open landscapes; in the early '90s it was the conservancy's largest conservation project to date. After Animas's purchase of the Gray Ranch (later renamed the Diamond A), the landscape remained encumbered by a conservation easement retained by TNC, which prohibited both partitioning of the ranch and any management practices that might result in significant environmental degradation.

Animas's goal of managing the property as a working ranch was a significant departure from the more mainstream and protectionist approaches to conservation. It returned the land to the hands of ranchers and sought to demonstrate that such enterprises could be both economically and ecologically viable. It was clear that the old paradigm of "bucks for acres"—of merely purchasing land to protect it—which had been TNC's mantra for the previous decades, had been pursued as far as it could go. Conservation science now demonstrated that to successfully conserve biota over the long haul, one needed to preserve large intact and unfragmented landscapes. TNC's experience in the developing world demonstrated that for large-scale conservation to be successful and sustainable over time, its proponents needed to respect and account for the needs of local people by actively seeking their support and collaboration.

In addition to maintaining ranching activities, an equally important focus of the ranch would be careful, scientific research and monitoring to document the ecological impacts of livestock. This work would serve as an example for future conservation efforts throughout the borderlands and across the West. Toward that end, the Animas Foundation hired senior conservancy scientist and conservation expert Ben Brown to oversee operations, and even proposed endowing a chair at the University of Arizona to support research programs on the ranch. Ben had worked with TNC establishing programs all over the West, including at the pre–Animas Gray Ranch. Seeking out his experience and university partnerships indicated just how seriously the Hadleys took their vision of science-based stewardship; their efforts introduced a conservation science–based framework to the region.

The Malpai Borderlands Group

Two very different, yet equally influential events led to the formation of the Malpai Bordlerlands Group (MBG), which would come to redefine the role of local ranchers from resource users to landscape stewards across the West. At the same time Drum Hadley was leading efforts to bridge the divide between conservationists and cattle ranchers, Quaker peace activist Jim Corbett was leading refugees from Central American armed conflicts across the Mexico-U.S. border through the region's rugged mountains. Corbett was a founder of the Sanctuary Movement, a religious and political initiative that sought to provide safe haven for refugees in the United States. Through Corbett, via his trek through the borderlands, local ranchers were introduced to Quaker ideals of consensus building and common ground. A small group of ranchers and friends, at the urging of Drum Hadley, began to gather regularly on the porch of Warner and Wendy Glenn's home on the Malpai Ranch east of the town of Douglas, Arizona, to discuss these ideas as they applied to the rangelands conflict. In the words of neighbor Bill McDonald: "We had got extremely good at knowing what we were against, we needed to decide what we were for." The porch discussions with a diversity of ranchers, researchers, and conservationists provided the essential forum for allowing the group to chart a new course.

In addition to playing host to philosophical conversations, the Malpai Ranch was also the setting of a more visceral event, which catalyzed the MBG: a wildfire that began on the property. Recognizing the potential ecological benefits of fire in reducing woody shrubs and helping to restore grasslands, the Glenns requested that local agencies not put it out. However, federal agencies were mandated to undertake fire suppression using the "ten o'clock" policy, which requires all fires to be extinguished by noon the next day. Debate ensued, but before any action could be taken, the Malpai fire went out on its own. Still, that experience, after many similar ones involving governmental entanglements, upset the ranchers' frugal sensibilities. They viewed both the time and expense involved in many fire suppression efforts as extremely wasteful. In the words of rancher Bill Miller, agencies "spent hundreds of thousands of dollars to protect structures worth a few thousand dollars, to eliminate fire that could be doing incalculable amounts of good."

When the Animas Foundation purchased the Gray Ranch, local ranchers were introduced to John Cook, a former director of TNC's Florida chapter, who had been tasked with finding a conservation-minded owner for the property. Like the legendary George Martin was to the Beatles, John essentially acted as the MBG's producer, crafting the group's mission and message, working behind the scenes with other TNC leaders, such as chief council Mike Dennis, public relations expert Kelly Cash, and local ecologist Peter Warren to make its vision a reality with the help of TNC's considerable financial and technological resources. The importance of Cook's and TNC's leadership to the development of rancher-science partnerships in the borderlands cannot be understated. Cook's immense energy and interpersonal skills were especially crucial in helping the ranchers get their organization off the ground.

Thus, the Malpai Borderlands Group was formally created as a nonprofit corporation in 1994, as a collaborative effort by borderlands ranchers to protect both their wild landscape and their way of life. The Animas Foundation, as a member of the organization, played a crucial role in getting the fledgling MBG off the ground, donating computers, geographic information system software, and financial resources to ensure that the group had the contemporary technology and organization to conduct effective, science-based stewardship. John Cook, along with rancher Bill McDonald, assumed the organization's co-leadership.

One of the first orders of business for the MBG was a collaboratively developed, landscape-level map of the region designed to guide coordinated regional fire planning (fig. 2.2). This fire map remains the primary tool for defining the extent of fire that complements the management strategy of each landowner each season. The map is still an important, tangible example of how ranchers and agencies can work together to craft commonsense solutions.

As it turns out, fire is a brilliant process around which to build a conservation program in the Southwest because it integrates all manner of ecological and social variables. Periodic fire is important for sustaining a diversity of habitats in rangeland ecosystems.

On the human side, fire brings together an array of public and private partners ranging from university researchers to local fire departments. The trust and other elements of social capital that emerged from the MBG's process promoted other facets of conservation in the borderlands, ranging from protection of endangered species to habitat restoration. Furthermore, fire also provides a powerful metric for assessing the threat of landscape fragmentation: if fire is essential to sustaining landscape-level processes, then anything that prevents large-scale fire is a threat not only to the land, but to its human communities and culture. Ranchers and their agency collaborators know from experiences in adjoining mountain ranges that even a limited number of inholdings with structures (e.g., vacation homes) renders large-scale fire management both ineffective and unaffordable. Managing fire to avoid harm to private property is extremely expensive, and cost per acre is the major variable influencing the ability to restore fire to the landscape. Because even a small number of additional structures can exponentially increase the logistical and legal headaches associated with prescribed burns, structures have huge implications for the viability of the landscape as a whole. For this reason, the purchase of conservation easements that limited subdivisions became

Figure 2.2. Contrast of vegetation types: An unburned area where piñon and juniper have reached high densities with little forage or ground cover and often high erosion, because the ground layer has nothing to impede water flow (left). The almost monoculture of vegetation simplifies the ecosystem, resulting in reductions in diversity and ecological function. A habitat burned seven years earlier (right). The grass not only provides forage for livestock and wildlife, but also reduces erosion. The mixture of habitats, from trees in the draws to grassland and savanna, generates much more landscape-level diversity and preserves ecological function.

routine for MBG. The borderlands fire program remains a rare example of collaborative and adaptive resource management that successfully returned fire to the landscape.

Fire works as a pivotal management tool because in the borderlands nearly a century of fire suppression, combined with the residual effects of overgrazing and the emergent impacts of climate change, leads to increasing rates of range degradation as woody shrubs and dense forest replace grasses.[3] Ranchers in the borderlands noted that in areas where fires were able to run their course and eliminate woody shrubs, once-prevalent native grasses returned. To explore the then-counterintuitive proposition that fire might actually restore desert rangelands rather than destroy them, MBG ranchers sponsored conferences in Tucson, in collaboration with TNC and federal agencies such as the Forest Service, on the effects of fire on the region's biota. Based on the bulk of scientific evidence, they concluded that only reintroduction of large-scale ecosystem processes such as fire could mitigate the impacts of dramatic vegetation change. Some members of the environmental community accused both the ranchers and agencies (and TNC) of simply seeking more forage for cattle, but missed the larger ecological implications for land health. Overly dense forests are prone to soil erosion and declines in biological diversity. Conservationists and scientists had (and continue to have) common purpose with the cattlemen over desertification in desert grasslands, an ecosystem imperiled all over the globe. In this sense, the MBG has essentially served as one vast climate change mitigation project, in which fire and grazing were used to mitigate climatically driven vegetation change (fig. 2.3).[4]

However, environmental threats were not the only challenges facing the group. In the years following the MBG's formation, rural subdivision and associated changes in real estate values and rural demography became major threats to both land and local culture. In the face of these pressures, positive interactions between scientists and the rest of the community became increasingly important for sustaining human and natural systems. After demonstrating common ground between ranching and environmental sustainability, the MBG's new task was to demonstrate how ranching was vastly preferable to exurban development. Experimental studies in the borderlands, as well as related studies on the effects of subdivisions elsewhere in the West,[5] supported the community's perception that without active management to preserve or restore natural environmental processes (such as fire), both traditional pastoral livelihoods and the characteristic composition of the land itself could be lost.

The Malpai group realized early on that the new kinds of problems facing the borderlands required a different kind of science than the range ecology that for decades helped landowners maintain livestock, wildlife habitat, and livestock forage.[6] Accordingly, the MBG sought the involvement of researchers who were focused on conserving dynamic systems and large open spaces, even if they did not necessarily have a history of working with ranchers.[7] They weren't looking for converts to a rancher's way of thinking; all they asked was that scientists come with their minds open to innovative responses to the challenges facing the region. Under

Figure 2.3. The driving variables addressed in the Malpai borderlands. Fire is the most cost-effective means of restoring large areas of landscape by removing trees and shrubs and restoring the patchy vegetation composition that increases biodiversity (left). Research in the borderlands (Curtin and Brown, 2001; Curtin, 2008) demonstrated that grazing (right) at moderate levels reduced climatically driven vegetation change and, like fire, increased vegetation patch dynamics and landscape-level diversity while also sustaining local economies and culture.

the guidance of board members such as Raymond Turner (arguably the dean of southwestern botanists, who had recently retired from the U.S. Geological Survey), and John Cook, Malpai assembled a team of creative problem solvers to rethink the role of science in conservation. Notable among these was the University of New Mexico's James H. Brown, one of the nation's most eminent ecologists. Brown instilled in the group an appreciation for experimental approaches to conservation. However, he pointed out that senior biologists, such as himself, were overcommitted, and suggested that the MBG attract more junior researchers, who could focus their careers on developing science programs in the borderlands and give the work the attention it deserved.

At the time, I was completing a postdoc with Brown's group at the

University of New Mexico. Because of my background in landscape and large-scale ecology, and interest in climate and land-use interactions, I was a natural fit. I was asked to take on the work of initiating a coordinated science program through my recently formed research institute, the Arid Lands Project. As I acquired a leadership position in designing the overall Malpai science program in 1998, the focus of my work underwent a radical shift from landscapes of the Intermountain West to the varied topography of the borderlands. This opportunity came with a steep learning curve, and profoundly influenced my conception of environmental challenges. Through collaboration with the ranching community and conservation professionals such as Cook, it was like going to graduate school all over again (only more intense). The development of the MBG was a crash course in conservation design and collaborative process that was to prove essential for designing a post-normal approach to science that embodied social and ecological elements of change.

The actions of the Malpai Borderlands Group are well chronicled.[8] The group has been immensely successful in restoring fire, acquiring and maintaining conservation easements, establishing adaptive management, and developing conservation plans for rare and endangered species. They also serve as a powerful, though often reluctant, model for ranchers and conservationists across the West, as well as in East Africa, Mongolia, the Middle East, and beyond.[9] However, much less has been explored regarding the implications of a foundation in post-normal science that provides the crucial informational and experimental feedback loops to inform and guide conservation policies and on-the-ground action. For the science programs were essentially a social, as well as an ecological, experiment. The insights proved to be equally profound in what they reveal about design for ecological and social resilience, as they were about the science itself. For these reasons the successes and the failures of the Malpai science program, at the time the largest on the continent, are chronicled in the coming pages, for they provide essential insights into how knowledge is gathered and attained in large and complex systems. The science experience provides a more detailed look at the social and political dynamics behind not just the research and monitoring, but borderlands conservation in general.

Designing Science for Open Spaces

During my many hours of driving to and from field sites in the borderlands, I had the opportunity to reflect not only on the beauty of these landscapes, but on the need for science- and community-based conservation to be conducted at scales that were directly relevant to conservation and management. This meant designing a process that embodies an appreciation for elegant experimental design as imparted by my postdoc experience managing the Portal Project, which was a highly replicated ecological experiment in the Chihuahuan Desert of southeast Arizona, and taking a dynamic, complexity-driven systems approach to science and conservation.[10] The key was synthesis—not just conducting haphazard, isolated studies and attempting to string them together after the fact, but a coordinated, carefully planned, and innovative series of studies that addressed a wide variety of environmental factors and ultimately served the needs of a diverse group of regional stakeholders, while also doing cutting-edge science. A key point was to show that far from being the impediment to scholarship that conservation and community engagement are usually perceived to be, they are actually complementary and even necessary to developing the kind of post-normal science essential for dealing with large, complex, and "wicked" problems.[11]

The core challenge then, as now, was that most institutional frameworks are simply inadequate for developing relevant large-scale conservation and science, with most studies too small and short-term to capture the underlying processes,[12] or ill-equipped to handle the multifaceted nature of environmental problems in landscapes as vast and fragile as the borderlands. The debate over grazing on public lands made it apparent that new institutions with pioneering organization needed to become reality fast. Without appropriately scaled and peer-reviewed studies on actual grazed systems, both ranchers and conservationists would continue to draw their own conclusions based on their respective biases and assumptions,[13] and the work of protecting the borderlands and the West for both interests would remain compromised and gridlocked.

But how was the Malpai Borderlands Group, even with the best of intentions and support of scientific minds, supposed to distill a million-plus-acre ecosystem with a seemingly infinite number of variables down into a coherent framework for action and research? It began by initiating a number of meetings between ranchers and researchers to better understand the ecological processes affecting the borderlands, while I spent hours in the saddle getting to know the land and the people. Of the many important variables affecting ecological integrity, we concluded that only two are readily accessible to management or experimental manipulation: fire and grazing (fig. 2.4). This is a critical point, for if a variable cannot be measured or manipulated, it is not useful from a science or management perspective. Fire and grazing, along with the overarching effects of climate, became our core monitoring and research priorities.[14]

To understand the impact of fire and grazing at large scales, the MBG established more than two hundred monitoring plots to assess plant community composition and diversity as indicators of ecological condition across numerous ranches to document vegetation change.[15] However, we soon realized that although our monitoring sampled a broad area, it was also largely uncoordinated and unreplicated (e.g., no control plots and some key variables such as rainfall were not assessed). Therefore, although effective at detecting patterns of change, it was poorly suited to determining their cause. Experimental studies that controlled for factors such as grazing or rainfall to complement the monitoring were needed to assess the outcome of climate and land-use interactions. Special concerns included how fire influenced biodiversity and ecological function, the role of climate change, and whether ranching could be compatible with conservation. As Bill McDonald (fig. 2.5) stated, "If ranching really is not sustainable out here, I would rather we be the first to know about it."

To honor its founding commitment to peer-review-quality science,[16] the MBG developed experimental approaches unparalleled by any other place-based collaborative on the continent.[17] This commitment to high quality and frequently experimental science was the cornerstone of the group's approach, which set it apart in the eyes of not just collaborators, but also agencies and funders.[18] This wasn't just good science; in the politically charged atmosphere of the 1990s, it was also an important survival strategy. Rather than participate in the hyperbole and rhetoric typified by both sides of the rangeland debate, with ranchers and environmentalists lobbing accusations at each other without seeking solutions to their common concerns, Malpai undertook collaborative, observable action on an

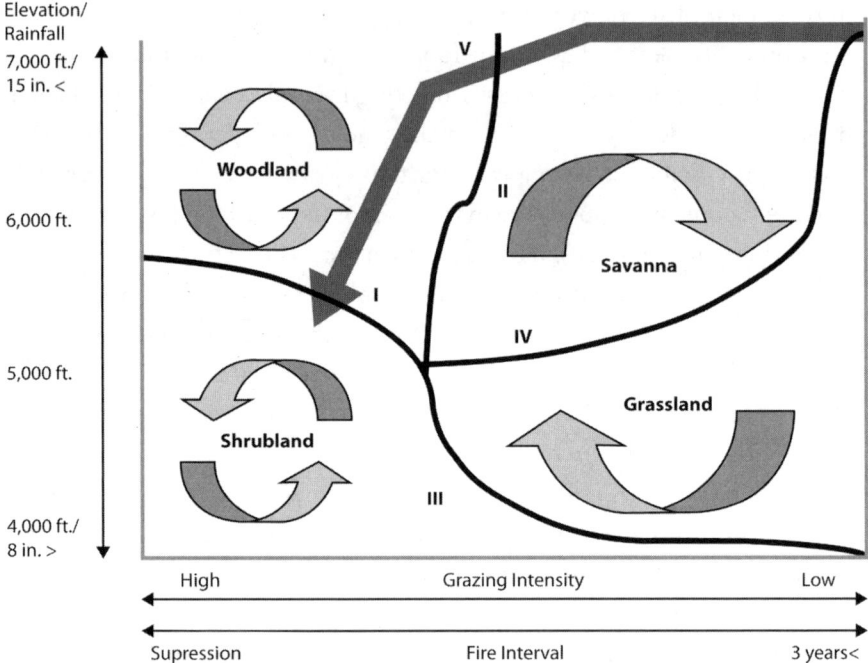

Figure 2.4. *This systems model of the Malpai borderlands makes two simplifying assumptions based on recognized environmental patterns: (1) rainfall and elevation are correlated, with increases in elevation associated with increases in precipitation, and (2) grazing and fire are inversely related, with high grazing intensities resulting in fire suppression. The Roman numerals I–IV bound the area encompassed by the borderlands research program. The downward arrow, Roman numeral V, indicates overall system decline and a loss of ecological or economic options; the actual location of this line and the corresponding thresholds of change in ecosystem function are unknown and were a key focus of borderlands research and monitoring. (Adapted from Curtin 2005, 2008.)*

unprecedented scale. This willingness to engage both sides gave them the credibility to undertake other forms of land conservation. The Malpai approach might best be summed up by John Cook's conviction, "Live by the sword, die by the sword." Invest in the best peer-review-quality science and "let the chips fall where they may," meaning that the MBG would, for better or worse, be science-driven and would abide by the results of the research and monitoring.

Figure 2.5. *Borderlands ranchers Bill McDonald (left) and Warner Glenn (right) played critical roles in forming and leading the Malpai Borderlands Group. Their leadership exemplifies the importance of having the local community involved in on-the-ground conservation action.*

Opportunities and Constraints

However, the borderlands science was not just about settling a political debate over pragmatic resource management and range science; it was intrinsically subversive in being aimed at remaking the way we do science by embracing a post-normal approach.[19] Landscape-level processes do not exist in isolation, nor are they managed in isolation, yet they are normally studied in isolation. Understanding these processes requires examining them concurrently at scales directly relevant to conservation and management. But this requires reconceptualizing fundamental approaches to research.

Most studies avoid, rather than embrace, complexity; complexity reduces the ability to get clear and reproducible results, while greatly increasing research cost and time.[20] And yet, recent studies have shown that the time-honored approach of working at small scales in microcosm studies and assuming their characteristics are representative of large systems does not work.[21] Large systems have fundamentally different properties than smaller ones. Just as children are not simply miniature versions of adults, a tree is not a forest. Understanding the workings of large systems requires embracing their inherent and ever-present complexity, as well as the emergent outcomes of interactions among diverse variables. This may dramatically increase the difficulty of getting statistically significant results due to increasingly messy and variable data; however, when patterns are detected, they stand a much higher chance of being more than statistical artifacts.[22] In a science of open spaces, emergent outcomes of myriad factors mean that results are relevant not just for theory, but also for meeting the pragmatic needs of conservationists, land users, and resource managers, while testing the limits of complexity-based approaches by using them to integrate theory and practice.

Such approaches may seem straightforward and obvious, but even relatively straightforward questions, such as "What is the environmental impact of grazing?," become much more complicated at the scales in which they exist in reality, rather than ecology's typical small-scale, isolated study plots. A typical grazing study design, for example, measures the impacts of cattle on a landscape, then removes the cows from some or all of it. The difference between the grazed and ungrazed landscape is determined and voilà, one has a measure of grazing impacts on vegetation or other biotic and abiotic factors. However, although the simplicity of this type of design is intuitively satisfying, it does not produce a measure of grazing impacts at all, but rather the lack of them. Although how systems recover from overgrazing is an interesting and important question for conservationists and land managers, it does not address how cows or other grazers actually interact with landscapes. To answer this question requires observing real, rather than hypothetical, ranching situations, which vastly increases the scope of the study.

To capture the complexity and dynamism of interactions that occur on the landscape, research design must itself be complex and dynamic, as well as adaptive. Understanding the effects of grazing on plant or animal composition or ecosystem processes such as soil erosion should therefore include not only cattle, but also native grazers such as bison or pronghorn. Fire too must be considered and studied appropriately, because it has dramatic implications for the outcome of grazing (and vice versa). All of these elements have synergistic and interactive effects; none exist in isolation. Of course, this means that the size of the landscape required for meaningful study of these processes must be, almost by definition, extremely large.

In addition, the larger the scale, the more social impacts must be accounted for. The social implications of research design mean that indirect effects of the research process itself need to be considered just as much as experimentation and replication. To understand the role of climate and landscape-level ecological and cultural resilience, the MBG had to meet several essential design parameters:

- To mimic natural occurrences, burns had to be introduced during the warm season, when fires normally occur, and at large enough scales to match local fire events.

- Grazing had to be practiced at large enough scales to reflect the actual management of local ranches.

- Our research itself needed to be conducted at large enough scales to investigate the interactions among core variables, including climate, fire, and grazing, that were key in shaping local ecosystems.

To meet these requirements, we needed access to substantial land-scapes with accompanying fencing and water, assistance with prescribed burns, and hundreds of cattle and the cowboys to manage them. By that token, our research design required maintaining and embracing local social, as well as ecological, complexity.

The nature of our design parameters meant the research could not occur on most federal lands, as has been the norm for most other long-term ecological studies, because of two factors: access and landscape-use history. In terms of access, fire management on many federal lands is usually highly restricted. Typically, fire is allowed only under the safest of conditions, which usually means burning on cool or windless days. However, these are conditions when fire almost never occurs naturally. We needed a study site where, on hot, windy, summer days, we could really let the fire run, generating the kinds of dramatic fire behaviors that exist in nature. We also needed a place that allowed extensive grazing, whereas many federal lands set aside for research explicitly do not allow grazing or do not have room for the kinds of large-scale grazing studies that mimic actual management.

The other issue with federal lands in the Southwest is that they consist, by and large, of areas abandoned by pioneering homesteaders, which over time have reverted back to public ownership. Productive, well-watered land historically remained in private hands. This means that in selecting federal lands for long-term ecological studies, such as for the National Science Foundation's Long-Term Ecological Research Program, one is often inadvertently selecting fragile lands that are often not representative of active ranches (or functioning ecological systems in general). Studies on degraded lands that are now desert but were once grassland can result in what may be intended as studies of desert, but are actually studies of desertification. As with the issue of grazing versus rest, the impact of desertification is an intrinsically interesting question, but it is a problem when the function of degraded landscapes accidently becomes a proxy for healthy ones. This concern was confirmed by Daniel Milchunas's exhaustive 2006 review of the grazing literature from the Southwest, in which nearly every study was conducted on degraded or brittle federal lands, illustrating that essentially all conventional knowledge from the region was drawn from heavily modified landscapes.

In short, we needed to find a place where historic records demonstrated that the ecosystem was essentially intact (fig. 2.6)—a place that was large, that allowed grazing and summer burns, and that was closed to outside human interference. Finding such an area seemed a tall order.

Figure 2.6. *Historical images of Mexico-U.S. boundary markers at the south edge of the McKinney Flats research area from 1893 (top) and 1994 (bottom), taken roughly a century apart, illustrate that the study site was relatively unchanged since European settlement, with much the same vegetation as existed in the 1800s. (Photos courtesy of Raymond Turner.)*

Luckily, the borderlands had just the right place. On the vast Gray Ranch, recently purchased by the nonprofit Animas Foundation and set up explicitly for conservation and research purposes, was a remote and little used pasture that perfectly fit the bill for developing the kinds of long-term, large-scale studies needed to help the Malpai ranchers and to understand southwestern ecosystems in general.[23]

McKinney Flats

In 1998, at the invitation of Animas Foundation's board, a multidecade experimental research program was initiated by the Malpai Borderlands Group on the 8,800-acre McKinney Flats pasture, located at a transition between arid grassland, shrubland, and savanna, with its southern border abutting Mexico. This project, which I primarily designed and directed along with the crucial help of researchers such as University of New Mexico's David Lightfoot, and independent range consultant Myles Traphagen, became the largest replicated terrestrial ecological experiment on the continent, examining complexity by focusing on the interaction between biotic and abiotic, as well as social and ecological variables.

Our research design essentially placed the group's collaboratively developed model of ecosystem change (see fig. 2.4) directly on the landscape to test it, explore its implications for conservation, and refine the parameters to better understand the thresholds of change in arid and semi-arid grasslands. This was key in allowing the group to quantify and test its assumptions about the role of ranching in being sustainable and beneficial to large-scale grassland conservation.

The McKinney Flats project was a critical departure from traditional ecological research or rangelands science studies in that it was intended to mimic actual ranching as closely as possible. Although traditional field experiments can achieve high levels of statistical precision, they are often unable to accurately depict the outcomes of unforeseen management actions. The irregularity of natural events such as drought and fire, the unevenness of grazing, and unexpected events, including water system failure and broken fences, all throw wrenches into the clockwork world envisioned by conventional experimental design. Yet these factors are integral for science to consider if it is to reflect reality in meaningful and consequential ways.

For example, the decisions cattle ranchers make regarding the size of their herds are difficult to predict—and therefore to replicate—in traditional field experiments. Ranch managers overshoot and undershoot desired stocking densities due to climatic variation, beef market fluctuations, time lags in stocking up the pasture, water access, and countless other variables. This increased variability generates more heterogeneity in the environment than can be attained through precise, rigid research protocols.[24] It also contributes to landscape-level ecological diversity, and potentially to greater overall landscape resilience in response to grazing and climate, for ironically it may actually mimic the patchiness of native grazers that were historically on the landscape more closely than the conventional target of grazing evenly and striving to consistently "take half and leave half" of the forage. With native grazers such as bison, through time their grazing is more irregular across the landscape, with some areas much more heavily used than many range scientists or managers would consider optimal, and other areas relatively unused. If studies are to be designed to accurately reflect the complicated effects of ranching practices, with the long-term goal of finding solutions to ecological degradation, they require new approaches that take such complexity and historical variability of natural grazing systems into account. McKinney Flats was intended to do just that.

But it was also crucial in providing empirical support for ranching as a conservation strategy by demonstrating the impacts of livestock on rangelands at landscape scales.[25] This outcome has been especially important in answering the charges of critics and skeptics who believe that all human-driven activity on the landscape is detrimental. This is because one of the challenges to changing public perception about ranching is that well-managed grazing is nearly invisible to the untrained eye. Unless the cows happen to be present, most people would be unable to tell a well-managed, grazed pasture from ungrazed grasslands. Overgrazing, on the other hand, is extremely obvious and tends to occur near riparian corridors or ranch entrances that people drive through or recreate in. Therefore, many people believe that most grazed lands are overgrazed because that is all they see.

Much of the debate is also an issue of scale. At local levels grazing can be extremely damaging if livestock are confined to small spaces. Yet at large scales, where cattle have room to move across the landscape, it can contribute to ecosystem diversity and function. So McKinney Flats tested at large scales the MBG's underlying premise that ranching contributed to, or at a minimum did not intrinsically conflict with, conservation.[26]

McKinney's experimental research and related studies (fig. 2.7) also changed perceptions about Southwest ecosystems by showing that large native grazers such as pronghorn perform a far more important role than previously thought in structuring the vegetation of desert grasslands,[27] an insight that continues to have significant implications for land tenure and ecosystem health across vast regions of the West. This insight, and the results of studies of bison in Mexico just south of our study area, re-defined what we think of as "natural" processes, and how to best sustain them, by illustrating how grazing can be important for maintaining eco-logical function. Ranching and livestock grazing thus move from being thought of as an exotic intrusion to part of a process intrinsic to main-taining ecosystem health by helping with nutrient cycling and maintain-ing a dynamic patchwork of different vegetation across the landscape.[28]

Our studies on fire in the borderlands, as well as companion studies conducted at the Jornada Experimental Range in southeast New Mex-ico, helped dispel the myth that fire is intrinsically damaging in desert grasslands, and resulted in more extensive use of fire as a management and restoration tool throughout the region. Conventional wisdom at the time held that fire was destructive to desert grasslands, although this misperception was based primarily on a couple of very brief studies conducted on desertified ecosystems during drought in the early 1960s (another example of assuming the results from stressed systems are in-dicative of rangelands in general).[29] In functioning systems that were not stressed, fire had no effect or even contributed to diversity and ecological function.[30]

In addition to addressing the ranchers' interests and putting the MBG's efforts on solid empirical footing, the most far-reaching insights from McKinney Flats emerged from research on complex interactions between biotic and abiotic variables, including climate, fire, and grazing, and how they influenced multiple guilds of organisms ranging from ants to liz-ards.[31] Located at the boundary of three different ecosystems (including grassland, shrubland, and savanna), the study allowed for examination

Figure 2.7. *Aerial image of McKinney Flats at an elevation of approximately 5,397 feet looking east across the study area. The perpendicular lines in the foreground are a cattle exclosure plot. The research area spanned a grassland/ shrubland transition zone, with higher levels of shrubs to the west and a desert grassland to the east. In also incorporating a savanna through oak stringers in the riparian zones and on the hillsides, the site contained almost all the midelevation habitat diversity of the borderlands and thus was a good model system for addressing questions of vegetation change in response to climate and disturbance by essentially placing on the ground the four-box model shown in figure 2.4.*

of the thresholds of change both spatially and temporally by examining climate disturbance interactions across the research pasture over many years.

The results showed the value of complexity-driven approaches in demonstrating that the interplay among variables was more significant than that of the same variables viewed in isolation. For example, viewed from a traditional perspective, fire might not be considered an important driver of landscape change if it accounted for only 8 percent of the data's statistical variation for the season immediately following a prescribed burn. However, this interpretation could be wrong, for although fire might represent a small proportion of variation, it can have huge amplifying effects on other system processes. The same is true for grazing.

Consider: fire eliminates fine fuels composed of mostly fibrous, nutrient-poor grass or leaf litter that are left from previous growing seasons.

After a fire, subsequent regrowth of grass (usually of the same species, for these are often long-lived, fire-adapted perennial bunch grasses such as blue grama [*Bouteloua gracilis*]) provides a relatively richer source of nutrients that attracts grazers. Native and domesticated grazers then continue to select these regrowth sites over unburned areas. Increased nutrients, higher nutrient cycling, and newer biomass generate more nutritious forage, perpetuating the pattern. Measures of livestock use on McKinney Flats documented that cattle occupy recently burned sites eight times more often than surrounding areas that were not recently burned.[32] Even if cattle are excluded from burned areas by fencing or other measures, the residual impacts of fire on grassland productivity remain. For native grazers are also drawn to burned sites, grazing them at orders of magnitude more than unburned or ungrazed areas and perpetuating the cycle.[33]

These outcomes highlight not just the significance of interactions, but also the role of lag effects in ecological processes. Thus, the impact of grazing in response to burning causes patchiness in herbivory intensity, which is superimposed on the patchiness of the actual burn event, causing dramatic shifts in the landscape-level architecture of vegetation. Fire, therefore, initiates cycles of interaction that can still be detected years later, even when immediate postburn impacts (sampled following vegetation regrowth) may be negligible.

The same holds true for interactions involving prairie dogs, in which nutrient-rich forage resulting from their herbivory attracts larger grazers such as pronghorn and cattle. Therefore, the cleared ground observable around prairie dog colonies is not just the outcome of their foraging, but also of synergistic relationships with cattle and native grazers. This, in turn, has important implications for prairie dog populations, because the clearer ground reduces their mortality by allowing them to better detect the approach of predators and leads to population increases and expansion of the overall colony.[34]

The most important outcome of the McKinney Flats study was not any one specific result, but rather the process itself: using empirical ecological data to test the assertion that collaborative, place-based approaches to science result in fundamentally different insights than conventional models of research.[35] This process, essentially a field test of post-normal science, shows how social and ecological systems are integrally linked.

Recognizing and respecting these linkages is essential for developing the kind of dynamic and large-scale science needed to understand our rapidly changing world. The process of designing studies to test and measure emergent outcomes spanning ecological, social, and physical systems is at the heart of developing durable responses and adaptations to environmental change, and represents the foundation of a science of open spaces.

Cascabel Watershed Studies

Following the development of the McKinney Flats project, a complementary watershed study was established in 1999 on national forest allotments on the Cascabel Ranch (fig. 2.8) abutting the Gray Ranch (about thirty miles northwest of the Mckinney Flats study).

The study focused on watershed-level responses by comparing erosion and vegetation changes on twelve paired watersheds (four burned in the summer, four burned in the winter, and four controls). Although conceived jointly with a range of agencies, ranchers, and scientists, this is where the structural similarities between the two projects ended.

Figure 2.8. The Cascabel study site in the foothills of the Peloncillo Mountains. At approximately 1,000 feet higher than McKinney Flats, it contains a savanna habitat type, in contrast to the grassland and shrub ecosystems that dominate McKinney Flats.

The Cascabel study was conducted following the traditional, hierarchical federal model, with a culture of top-down direction rather than the collaboration typified by McKinney Flats. As such, it generated opportunities for contrasts of the two projects' very different approaches to science (a focus on single variables versus complex interactions), as well as how their opposing governance structures influenced the acquisition of knowledge (conventional versus post-normal).

Even though its individual project leaders were well intentioned, Cascabel's goals suffered under the traditional system of directed science. Its hierarchical power structure and centralized decision making revealed the pitfalls of top-down governance processes in scientific studies in a number of ways. First, though relatively rich in financial resources, it did not take advantage of opportunities to leverage knowledge from other projects. The Forest Service researchers became increasingly insular, and the spirit of collaboration and integration with McKinney Flats and other studies in the borderlands soon vanished.[36] The lack of continuity and consistency meant the sampling protocols fluctuated from year to year, undermining the initial experimental design and disrupting the integrity of the data. Poor communication meant that some studies were repeated with new sampling plots sometimes placed right on top of existing studies.

Rather than improving sampling designs and seeking innovative forms of collaboration that would leverage borderlands science, the researchers were hamstrung by governance that seemed intent on "doing the wrong things, righter," rather than practicing self-assessment, allowing external critique, or embracing reflective approaches.[37] The pathologies of the hierarchical approach were also in evidence throughout the region. Increasingly limited financial resources for research were sucked in to facilitate redundant and frequently poorly conceived sampling programs that supported Forest Service–associated researchers, rather than solid peer-review-quality science as laid out in the founding principles of the MBG.

In the end, the more traditionally directed research approach simply did not have the institutional capacity to effectively collaborate and integrate with local place-based projects. The federal research that was supposed to contribute to the MBG vision instead drew financial or organizational resources away from the rest of the borderlands and compromised the entire science program. The ultimate irony is that despite the byzantine complexity of fire management and coordination in a federal system, prescribed fire intended only for the summer burn plots got away, charring many of the control plots and rendering much of the data collection—and years of investment—almost useless.[38]

In conclusion, a side-by-side comparison of conventional and post-normal research exhibits almost diametrically opposite outcomes. The Cascabel study, by becoming disassociated from the MBG community and local scientists, contributed little toward a broader understanding of borderlands ecosystems, while at the same time it diverted substantial resources. Conversely, the McKinney Flats project at a lower annual cost engaged the local community via summer internships and other programs that helped support and inform the local community. In addition, McKinney scientists worked tirelessly on behalf of conservation across the entire region, frequently communicating their results on behalf of the MBG both nationally and globally, often at their own expense. Even in terms of resulting publications, Cascabel produced only a handful, most of which were never peer-reviewed, often existing in the federal "gray literature." McKinney, with far fewer resources, produced a significant monograph in addition to numerous peer-reviewed papers, and continues to produce results to this day.[39]

The contrast is striking. The place-based process typified by McKinney in essence put in place a positive cycle of research and outreach in which every dollar did double or triple duty collecting data and engaging and employing the local community, while also communicating the group's broader goals and aspirations and providing a foundation for additional conservation and outreach. By contract, the Cascabel, and many other federally managed borderlands projects, became as much about consolidating resources as they were about addressing issues of landscape integrity. The primary result was the near-term employment of federal researchers, with few added benefits, profoundly illustrating the intrinsic limitations of the governance and incentive structure associated with conventional institutional designs.

An important facet of a proposed science of open spaces is that we reconsider the basic processes by which science is undertaken. The collaborative and integrative process of a post-normal framework illustrated by the McKinney Flats program is in considerable contrast to that of the federal system and many academic research paradigms,[40] highlighting that place-based science can contribute considerable additional benefits (such as directly informing the community through a student intern program) and can ultimately be more productive, while often working at larger scales, for less cost.

Outcomes of a Post-Normal Approach

Over more than a decade of collaboration on monitoring and research projects the science met its intended purpose, the Malpai ranching community became knowledge brokers, independently supporting research that could not be undertaken by state and federal agencies alone. The ranching community's participation in borderlands science projects gave them the credibility necessary to engage critics in the political arena, shifting the balance of power for control over their landscape away from government administrators and environmental groups in urban population centers and back to the local community. Wherever implemented, the MBG perspective encouraged a systems-based approach to conservation necessary for working in large, dynamic systems that was a natural extension of pastoralists' understanding of the world. This holistic focus on preserving ecosystem processes supported the idea that preservation of large landscapes is essential for conserving the culture as well as the ecological integrity of the region.[41] Through their dedication to these ideals, Malpai demonstrated that a different kind of approach to science, one that embraced community, collaboration, and complexity, could be especially powerful in making conservation more effective. By allowing the science to work at larger scales and better address the intersection of human and natural processes and the cross-scale dynamics essential for understanding how to sustain resilient ecosystems and institutions, the post-normal science typified by McKinney Flats and related studies was not only more robust in testing underlying theory, but also more relevant to the community's everyday life.

Social Realities and Economic Constraints

As with any process the Malpai science program was designed with certain implicit assumptions. Principal among these was that access to land

and funding would continue at least as long as the MBG did. With a minimum of sixteen to twenty years considered necessary to capture several El Niño/La Niña cycles and at least one phase of the Pacific Decadal Oscillation and the accompanying drought and rainfall patterns. But perhaps the most fundamental assumption was that success breeds success, that if we demonstrated significant results in five to seven years, the longer term continuity of the program would be ensured. However, all three of these assumptions turned out to be false. The experimental science program came to an abrupt end in 2010.

How and why the assumptions were wrong is the focus of the duration of this chapter, and provides crucial insights into how science succeeds and fails in large, complex systems. The borderlands experimental science program appeared to be a win-win all around. It gave the Animas Foundation and Malpai Borderlands Group credibility and increased understanding of the environment and opportunities for better management, while allowing scientists access to a vast ecologically and culturally relevant laboratory.

For the Cascabel the end was not unforeseen. The escaped controlled burn that had compromised the overall experimental design, and a new era of fiscal austerity in federal agencies, meant there were many fewer resources available. More dramatic was the Animas Foundation board's vote to discontinue access to McKinney Flats without justification, thereby ending that research program. This occurred despite the work having continued support from Malpai Group, and over the objections of many of the region's premier scientists. The U.S. Geological Survey's Dr. David Mattson wrote, "This work is also an exemplar of translating science into action and into terms otherwise particularly meaningful to land managers." The University of Arizona's Dr. Thomas Swetnam noted, "This body of work stands as an outstanding exemplar of the employment of world-class science in support of landscape-scale natural resource management." Notable in their inaction, however, were major conservation organizations, which had played such a pivotal role in establishing science in the borderlands, but which now abandoned support to avoid alienating the Animas Foundation. As one staffer admitted, "[the] complexity of our relationship with the Diamond A [Ranch] precluded support at this time."

The scuttling of McKinney Flats research program was a shock not just to the scientists, but also to a number of the MBG's leadership, who expressed dismay that a program they had supported for over a decade with more than million dollars was terminated without MBG board input. It was not just a breach of trust and violation of long-term commitments to neighbors and funders and the founding principles of the Malpai Group, it was just plain irrational given the Animas Foundation's long-term stated goals. McKinney Flats research had more than delivered on its promise of experimental science and community engagement. It was well documented in both scientific and popular media through articles ranging from *BioScience* to the *New York Times*. The MBG's commitment to peer-review science, as typified by McKinney Flats, had given them and the Animas Foundation a level of credibility enjoyed by few collaboratives, while it was clear that the program had only just begun to fulfill its promise. For the scientists involved, all these factors deepened the sense of loss over the program's end. As one long-term researcher commented, "I guess we bet on the wrong horse," sharing the general sense of disillusionment by the research community that had invested so much in making the program work.

However, the loss of the experimental science program was not an isolated event, but part of an across-the-board elimination of researcher access by the Animas Foundation, ending a one hundred–year tradition of scientific inquiry in one of the most biologically significant places on the continent. The irony was inescapable that a nonprofit foundation, specifically developed for conservation and science purposes through its purchase of the Gray Ranch and through subsidized loans from major conservation groups, ended up being the organization to pull the plug on not just experimental science, but also a century of natural history in the region.

However, the circumstances resulting in the program's end did not exist in isolation, but were a reflection of vast social forces, including increased political instability along the border and a general decline in support for place-based conservation across much of the West. In the borderlands, threats, break-ins, and the murder of rancher and Malpai board member Rob Krentz displaced longer term scientific concerns as ranchers and collaborators came to focus on more immediate and personal threats. In the meantime, as for the effects of climate change, the over-arching issue for which the Malpai science program had been designed, no other study or program has filled the void. Though no region is exempt from the threat of climate change, the borderlands are projected to be especially hard hit by shifts in rainfall timing and distribution, a major challenge now left largely unattended.[42] The twin specters of nationwide recession and border security, coupled with the social challenges inherent in maintaining any long-term conservation or science program, demonstrate the challenges of sustaining place-based collaborative stewardship and conservation long enough, and at scales large enough, to make a lasting difference and illustrate that, to be sustainable, a science of open spaces must include solid social and institutional design, as well as relevant research.

Lessons

Could this dissolution of Malpai's collaborative, experimental science-based approach have been averted? In a 2006 internal report, I concluded that without dramatic changes, peer-review-quality science in the border-lands would be impossible to sustain. The report noted several key problem areas, primarily that principles of openness and transparency, which were a cornerstone of the MBG, did not extend to the science community working on the Diamond A Ranch. Even when in the process of conducting research on behalf of the MBG, researchers faced antagonist behavior from ranch personnel, who became increasingly entrenched, with many of the more open-minded and supportive staff terminated or forced to leave, while individuals who were territorial and hostile to science became increasingly powerful (a point noted not just by researchers, but also by other members of the borderlands community). Beginning in about 2001 there was a mini program of "apartheid," in which researchers were quarantined in a separate part of the ranch and our movements restricted to only the research sites. This further deepened mistrust and misunderstanding and set in place a cycle of negativity that could not be broken, despite the research community's best efforts to reach out and expand contact with the ranch and some efforts to intercede on the part of members of the MBG leadership.[43]

As a result the science community was put in the untenable position of trying to meet Malpai's long-term commitment to science without the support of the Animas Foundation, the group's largest and wealthiest member. Although we understood that there were limits to what the MBG leadership could tell its neighbors (especially wealthy, powerful ones), most researchers deemed the conservation community to be complicit in the demise of experimental science programs. Without external influence encouraging the Animas Foundation to meet its commitment to science as spelled out in the organization's founding documents, board statements, the original Memorandum of Understanding for McKinney Flats, and even a web page highlighting the importance of science on the ranch (since removed), there was little the science community or MBG could do to change the course of events.[44]

The lack of external pressure seemed to embolden Animas, which ultimately removed all of the researchers' gear from the field station in complete violation of the MBG's principles of collaboration and fair play and a legally binding access agreement. The researchers had become a pawn in a game of power politics between conservation interests and the foundation, in which science and the careers of researchers who had committed more than a decade to helping the community and conserving the region were sacrificed for the sake of political expediency. However, the reality was that the conservation community and the MBG leadership both found themselves in a double bind when considering action against Animas. Although the goals of the foundation strayed increasingly far from their founding mission, alienating Animas and driving them further into a self-imposed isolation could have further compromised conservation in the region. The outcome of the Animas experience suggests that with conservation ranches, more than just agreements to periodically monitor and maintain cultural resources need to be in place—that agreements about institutional goals and capacity also need to be enforceable to be sustained.[45]

The shift in the ranch's relationship with researchers was in part a reflection of Animas employees having increasingly little foundation in biology or range science. A certain antiscience culture came to pervade the organization, one which was in considerable contrast to most large ranches in the West, where there are managers with extensive professional training and there is increasingly science-based management. After the ranch coordinator and chief scientist was let go in 2006, there wasn't a single employee left on the ranch with any advanced training in natural resource management. Crucial rainfall data were lost and other monitoring discontinued. In the span of a few years, the role of the Diamond A Ranch shifted from that of a testing ground for sustainable ranching and relevant science into what was essentially a private hunting reserve, with most vestiges of the brilliant vision that created the foundation largely gone. The few ongoing management actions mostly consisted of predator eradication programs intended to promote game species,[46] as rainfall and vegetation monitoring that are essential for effective land management declined to levels well below those of the previous owners, who had no commitment to conservation at all.[47]

Overall, the fate of the Malpai Borderlands Group's goal to sustain a peer-review-quality science program is instructive as a microcosm of larger issues facing place-based conservation. Though the idiosyncrasies of the Animas Foundation are anomalous, they point to the importance of sustaining founding principles of openness and respect for different constituencies involved in partnerships and holding everyone accountable to the principles of the organization. The heart of the problem of sustaining effective collaboration lies in the broader challenge of reconciling the very different perspectives of various players in the borderlands, and the critical impact even relatively subtle shifts in governance can have in keeping relationships healthy.

For example, though independent researchers played a key role in supporting the MBG's initial vision, after about 2000, they did not have a seat at the table in decision-making processes that directly affected them. Though the point was made that having science too closely associated with the MBG could undercut the group's overall credibility by making the science appear biased, at the same time, by making researchers the lesser of "equal" partners, critically important lines of communication that developed a sense of esprit de corps eventually broke down. Though we continued to have an extremely close working relationship with the Malpai leadership, and received strong, unwavering support from the group's executive director, Bill McDonald, my sense at the time, and today, is that the reduced connection with the ranchers allowed the Animas Foundation to isolate the researchers in a way that was not possible earlier in the group's evolution, when the science was more front and center in MBG's activities.[48]

Sustaining Collaboration

Collaboration is often like a card game in which everybody plays by a different set of rules, and like a card game, each group's assumptions, end goals, and values are often hidden from view. Cultivating and sustaining social elements of large-scale dynamic approaches to conservation and research often boils down to uniting participants from disparate institutional cultures who are motivated by different reward structures. For example, academics' focus is often on generating peer-reviewed papers, whereas for practitioners, program development may be a more significant goal. Scientists are often rewarded for taking chances and having research success, as federal managers are more often promoted on the basis of their not having made mistakes. Collaborative groups such as the MBG build common ground, but this only goes so far in diminishing the differences in perspective among the constituencies.

Early on in the development of the MBG, ecologist James H. Brown announced before an audience of ranchers that "we [as scientists] are not here to tell you what you want to hear, but what the data tell us." Jim's no-nonsense, forthright demeanor served as a model for me and other scientists as we sought to define our role in the larger community as individuals who could put politics aside and looked out for the long-term interests of both the landscape and the local community.

However, ours was a radically different philosophy from that of other players in the conservation game, whose main role was to unconditionally support, rather than to critique or occasionally challenge, the MBG's activities. In this complex social arena, the difference between nongovernmental organizations (NGOs) and agencies serving their constituencies' near-term interests, and experimental scientists, taking the long view, was profound. Given this social gradient science programs are inherently difficult to sustain despite their recognized long-term benefits. Although agencies, NGOs, and donors represent tangible and relatively immediate support in terms of financial or logistical assistance for local people, working with researchers is more problematic in that it requires giving up some measure of control for more amorphous and much longer term assets, such as credibility and ecological understanding. Similarly, once a research program is in place, it cannot be turned on or off like water from a tap, but requires continuity that is sometimes inconvenient and which can be disruptive to ranch operations. This means that if institutions sincerely want to sustain social and ecological learning despite sometimes challenging social and economic environments, they need to develop governance structures and policies that counteract the kinds of corrosive political forces that were the eventual undoing of the Malpai science program.

Like entropy in physical systems, in which entities need constant inputs of energy to remain "self-organized," social systems also require large inputs of time and resources to counteract the intrinsic movement toward disorder. The outcome of the Malpai experience illustrates that upon entering a collaborative process, scientists and landowners need to understand the inherent differences in each other's constraints and clearly communicate goals and purpose. Getting agreements in writing, so there is less opportunity for misunderstanding or disagreement later on, is also crucial. However, all parties need to recognize that sustaining collaboration is much like sustaining a marriage; it takes understanding, compromise, and a lot of work. In the Malpai example there was not just a failure to sustain the science, there was also a failure to implement the lessons learned from it. The reasons for this will be discussed in the next sections.

Sustaining Funding

Declining levels of funding constituted another factor in the demise of Malpai's experimental science program. Though the MBG faced shrinking grant opportunities in general after the standard five- to seven-year honeymoon period of most NGOs, the challenges were especially acute for experimental science after the economic downturn in 2008.[49] One contributing factor was a lack of sophistication by funders who, by and large, did not distinguish the difference between experimental and observational approaches to data collection. Time and again, we found monitoring results stemming from observational approaches to be at best unpublishable, and at worst almost useless; without coupling the results with data on rainfall or other core variables, it was impossible to know what the patterns of vegetation change indicated. Yet funders consistently requested and supported monitoring while being naively unaware of its limitations. As Bill McDonald once wryly commented, vegetation sampling was "just a really expensive way to measure rainfall." Nevertheless, Bill was forced every year to defend the value of experimental science to increasingly reluctant funders, while the monitoring was rarely questioned. The irony was that the experimental approach was what allowed Malpai ranchers to test their assumptions and generate new knowledge, and that, after a decade of establishment, research costs were declining due to greater efficiencies in procedures, while monitoring costs were continuing to grow without producing much in the way of usable results.[50]

Although it is impossible to generalize, at least in the case of the foundations I worked with, another key issue was that many of the bold and innovative senior program officers of the 1990s had retired or moved on and were replaced by more junior staff. The process frequently became almost inverted with grantee organizations often containing more experience than the organizations that funded them. The junior staff were frequently more conservative and risk-averse than their predecessors, reducing the transformative capacity of many foundations and their willingness to sustain longer term programs.

Another significant factor in the decline of the science program was change in access to federal funding for research. The longevity of the Malpai's experimental science program was largely predicated on reliable access to federal dollars. Yet, turnaround times for reimbursements after a few years expanded from ten to thirty days (the maximum allowed under federal law), then averaged sixty to ninety days a decade later (with some reimbursements taking as long as 120 days or more), eliminating the independent researcher's ability to maintain a balance of payments. The paperwork itself became more onerous as well, part of a process designed to stamp out corruption but which seemed more effective in stamping out small, efficient, and innovative NGOs. Cooperating agencies under increased federal scrutiny found it easier to simply fund themselves than deal with private organizations, which further broke down the incentives for collaboration. This example profoundly illustrates the huge indirect effect funding protocols and process have in dictating conservation outcomes.

Severed Feedback Loops

The story of the MBG science program is emblematic of larger challenges faced in developing sustainable place-based collaboratives, not just in the Southwest, but anywhere it is attempted. Just as soil type, rainfall amounts, and other climatic factors help determine what types of plant life can grow in a given ecosystem, social preconditions and governance structure are the bedrock and soil that give rise to, promote, constrain, and profoundly influence the effectiveness of organizations. Trade-offs between short-term (financial) and long-term (ecological integrity) interests point to the inherent tension of developing sustainable approaches in working communities, including the need for well-developed governance structures to prevent short-term financial constraints from trumping long-term conservation opportunities.

In the early days of the MBG, frequent interactions between ranchers and their collaborators meant that the two groups reached high levels of common ground in their concerns about the borderlands. However, the intensity of those interactions was not sustainable. Personal and working relationships ebb and flow, long-time residents move on while newcomers arrive, and common ground can fade over time. To address these social factors, questions were raised about how to institutionalize the science and conservation programs. Following the early, ad hoc science meetings developed around issues of fire and grazing, a science symposium held in 1999 took stock of existing research in the borderlands, attracting eminent researchers from around the continent and globe.[51] A science advisory team was set up, composed of experts from a range of social and natural sciences whose role was to advise, and, most importantly, to critique the MBG's efforts.

The science meetings became a yearly fixture that, to this day, draw researchers from across the continent; many still consider it their favorite professional gathering of the year. However, as the meetings became more institutionalized, they lost much of their original focus on problem solving and critical analysis. Though the research itself was peer-reviewed, there was little critical external review of the science or conservation programs themselves.[52] This ultimately hampered prioritization of goals, resulting in a protracted tug-of-war between scientists trying to meet their commitment to inform the MBG's actions and the overall social dynamic, which increasingly excluded them from the decision-making process.[53] As such, we often found ourselves taking ethical stands that met the long-term interests of the MBG and the land, at the cost of our near-term standing in the community.[54] As the MBG became more widely recognized and successful, the number of individuals within our ranks willing to tell the MBG (or Animas Foundation) what they wanted to hear grew substantially; these interests increasingly displaced effective critique and longer term goal setting envisioned at the program's inception.

These experiences illustrate the importance of sustaining not just yearly meetings of science advisors, but also of developing institutional structures that regularly critique and evaluate the conservation and science process and feed that information back into science and management and to funders. Developing institutions that hold people accountable and reward effective programs or cut, reorganize, or mitigate poor ones, is crucial. In particular, regular input from a panel of senior scientists with no direct investment in the borderlands (as many organizations receive) would have benefited the MBG in a number of significant ways.[55]

First, it would have given the MBG more leverage with its agency partners to maintain research and monitoring efforts that hewed more closely to the jointly developed goals established in the 1990s.[56] Because federal grants and agreements are by nature political, external advisors could have given the group the leverage it needed to push the Forest Service's Rocky Mountain Research Center to more effectively integrate their work with the goals of the MBG. Instead, the Forest Service claimed it was generating benefits for the local community while, as discussed earlier, its research became increasingly self-serving, not just at Cascabel, but across the borderlands. Without external oversight, scarce resources were increasingly allocated to agency researchers' pet projects, or to poorly designed and implemented research that did not meet the MBG's standards of relevant scale and peer-review quality.

Second, external review would have provided more opportunities for overall goal setting, and equally important, a chance to generate better synergy among different parts of the program. As it was, monitoring and experimental science never complemented each other to the extent possible, especially as the agencies reverted to old habits of expert-driven science and diminished integration with the community and with other local projects.

Third, declining levels of interaction between researchers and ranchers reduced the utility of the science program. Over time, researchers lost the ranchers' regular input into the research process, as well as the means to allow science to better inform ranching activities. Thus, the feedback loops that allowed the science to provide a relatively objective counterpoint to political pressures were mostly severed. Turnover on the MBG board meant that many new members had not been a part of the original science-based goal-setting process or the battles over the roles of fire and grazing at the inception of the group that were won through a foundation in high-quality science, further deprioritizing strong researcher/rancher relationships going forward.

Finally, external review could have recognized the need to educate funders and to provide regular justification for financial priorities, especially because much of the funding community did not fully understand the utility of experimental science (especially that of a post-normal approach). Funders' priorities increasingly drove the research, rather than the priorities of the MBG or the scientists directly engaged in the research process. The ends justified the means, but soon the means came to dominate both the process and the long-term outcomes of the conservation and science.

Meanwhile, a long-term strategic plan commissioned in 2003 was supposed to have guided the policy process, but was undertaken by an anthropologist with little formal training in organizational design or resource management. Titled "Ecosystem Management in Conditions of Scientific Uncertainty," the report, though well written and often compelling, contained little of the critique and goal setting expected in such documents.[57] Peer review of the plan was almost nonexistent; when suggestions were made to improve the report, they were largely ignored. The strategic plan was an immense lost opportunity for what could have provided comprehensive guidance at a crucial juncture in the group's development. In the end it was largely a political document that told the community what they wanted to hear, while providing little or no comprehensive assessment or essential action steps on how to move forward.

Without a well-crafted strategic plan, MBG programs were left rudderless. Riddled with inaccuracies, the strategic plan actually moved the resource stewardship process backward, playing down the principles of science-based decision making upon which the MBG was founded and providing no guidance toward much-needed integration of research and monitoring. In the end, rather than being strategic, it led to ad hoc, reactive decision making.

The report was intended to be a living document—the first step of a long-term process of revision and improvement—but such changes never occurred. The document served only to institutionalize existing pathologies, for it did not promote the kind of inclusive goal setting, governance design, or critical review that was needed to sustain programs in the long haul. The lack of a clear expression of priorities and process sounded the death knell for sustainable science and adaptive management, because without formal guidance, policy loses its ability to attain stewardship goals and sustain existing programs (especially in the presence of the aforementioned tension between short- and long-term rewards and outcomes). This appears to have been what happened as the science-based approach envisioned for the MBG, and its commitment to peer-review-quality science, was supplanted by a more opportunistic and less coordinated approach. The outcome is a profound illustration of the importance of maintaining well-defined goals, true external review, and well-developed governance geared toward maintaining a viable process. The consequences when these preconditions are not met have implications not just for research, but also for the sustained effectiveness of the entire program.

Concluding Remarks

Collaborative approaches bring out the best and worst in people: courage, vision, and commitment, but also pettiness, shortsightedness, and greed. Which ultimately prevails is the outcome of developing formal and informal norms and principles that are fixed in values, yet flexible enough to allow adaptive process to respond to change and learn from experience.

The Malpai Group has been an icon of successful community-based stewardship and emblematic of what can happen when diverse groups set aside their differences to address a common goal. In this case the goal was to preserve working landscapes of the Mexico-U.S. borderlands through reintroduction of fire, conservation easements, peer-review-quality science, and collaborative adaptive natural resources management. It is difficult to overstate the MBG's immense achievements not just in the borderlands, but in redefining what is possible in preserving working landscapes across the West and beyond. This makes the experience of the experimental science program in attempting to challenge conventional science paradigms all the more striking in that it achieved its goals and more than delivered on its promise, but still was prematurely terminated. The critical question is why? How does even a best-case scenario such as the Malpai Borderlands Group lead to counterproductive outcomes?

Our experiment in post-normal science illustrated that conducting effective and relevant scientific research is largely an issue of scale, and that getting the scale right is as much a social challenge as a biological one. The Malpai Borderlands Group has made immense strides toward conserving its landscape and demonstrated that large-scale collaborative science can answer fundamentally new and different questions that can surpass conventional methodologies in understanding complex systems. But it also highlights the crucial importance of governance and process design. A science of open spaces integrates these lessons to explore the implications of developing more design-oriented paradigms for sustaining large, complex systems.

In the West, the word *governance* has negative connotations for its proximity to the term *government,* which for many conservative residents implies bureaucratic meddling and a loss of control. Yet it was clear from my experience in the borderlands that whether one calls it adaptive governance, policy design, or collective decision making, the ability to develop effective ways of gathering knowledge, learning from mistakes, and sustaining programs requires institutions that can translate peoples' goals and values into concrete, meaningful action.

Working in another large ecosystem—the Gulf of Maine in the western Atlantic—provided an opportunity for me to explore many of the questions of governance design raised by the Malpai experience in a completely different context. An examination of Maine's ocean fisheries, considered in the next chapter, provides insight into how decision processes can contribute to, or undermine, the social and ecological resilience of open spaces.

CHAPTER THREE

Experiments in the Governance of Maine's Coastal Fisheries

*Doesn't it seem to you ... that the mind moves more freely
in the presence of that boundless expanse, that the sight of it
elevates the soul and gives rise to thoughts of the infinite and
the ideal?*

               —*Gustave Flaubert,* Madame Bovary

Twenty-five-hundred miles northeast of the sky islands of the borderlands, off the coast of Maine, is another archipelago. Hundreds
of bays and inlets form a rich tapestry of different marine and terrestrial
habitats. Myriad rivers deliver nutrient-rich water to the sea from their
headwaters in mountains hundreds of miles away. Cold, phytoplankton-
filled currents traveling thousands of miles from the Arctic thread their
way through shallow banks of glacial sediments that guard the eastern
approaches to the Gulf of Maine. Some of the world's most extreme tides
expose and then submerge the intertidal zone twice every day. Meanwhile, vast gyres of currents circulate warm and cold water, distributing
nutrients and delivering lobster larvae and other species from the depths
of the ocean to the protection of nearshore waters (fig. 3.1).

In profile, the underwater basin and range topography of the nearshore region resembles the landscapes of the Malpai Borderlands. However, in contrast to the time-weathered borderlands, the Gulf of Maine
is a relatively young ecosystem, a product of the most recent ice age. In

Figure 3.1. *The Gulf of Maine in the northwest Atlantic Ocean off the northeast coast of the United States and southeast coast of Canada was one of the world's richest marine systems and fisheries. Like the borderlands, it is a crossroads of ecology and culture where decades of conservation and management effort provide fundamental insights into large-scale science and policy.*

the short span of a few thousand years, the Gulf of Maine became one of the world's richest marine ecosystems, its nutrient-laden waters once supporting the highest marine mammal diversity on the planet.[1] In addition, fish bones collected from pre-Columbian human settlements reveal a huge quantity and diversity of nearshore fish dating back thousands years. From canoes, Native Americans caught shark, swordfish, and vast numbers of large cod. Later, Europeans were drawn to the fringes of the New World to partake in the rich harvest of cod and other bottom-dwelling "groundfish." Many decades before the first permanent settlements in North America, European fishermen came to settle Monhegan Island, Richmond Island, and other locations in seasonal fish camps where they dried and salted fish for markets in Europe. This fishing heritage is the foundation of the communities and cultures that have lived on the coast of Maine historically and up to the present day.[2]

My work in marine systems came about through an effort to understand how governance principles could address shortcomings in rangelands conservation by providing more institutional continuity for decision making. In particular, I wanted to understand how institutional design influences the effectiveness of conservation and sustainable management of natural resources. Governance principles have been extensively applied and experimented within fisheries. The dramatic successes and abject failures provide invaluable examples to be applied to other systems. These questions were addressed through my work in helping to establish (1) the Downeast Initiative, a collaborative based not in a single community, as in the borderlands, but distributed across 7,500 square miles of the western Atlantic Ocean abutting the coast of Maine, (2) restoration of anadromous fish to Penobscot Bay through establishment of the North Island Science Cooperative, and (3) the Downeast Fisheries Partnership, which works to unite fishing communities in eastern Maine.

The emphasis of these projects was on diversification of coastal fisheries, as it became evident that the 400-year-old commercial fishing tradition in the Gulf of Maine was in trouble. What once was a vibrant industry and way of life built upon the harvesting of everything from cod and flounder to haddock and bluefin tuna is becoming increasingly reliant on a single species: the American lobster *(Homarus americanus)*[3] (fig. 3.2). No single fishery has ever been sustainable over the long haul, so for the majority of fishermen to narrowly focus on the same species represents a considerable risk to the economy and culture of the coast, while the lack of ecological diversity makes the entire system less resilient to environmental change.

Lobster once played a minor role in the working life of fishermen; boys would take lobster from skiffs, while fishermen in their prime would work primarily offshore for groundfish such as cod and haddock, only returning nearshore for lobster in old age or to offset other catches.[4] However, among recent generations of fishermen, it is easy to find individuals who have never fished for anything but lobster. The soaring numbers of lobsters caught each year have paralleled a rise in affluence of

Figure 3.2. *Lobstermen hauling their catch at dawn off the coast of Maine. Over time the boats have become larger, more specialized, and more expensive, further capitalizing a fleet that is likely to see dramatic changes in income if lobster populations ever return to lower long-term averages.*

lobstermen. New pickup trucks and fishing boats of ever-increasing size and horsepower are becoming common as the fishery becomes more specialized and reliant on a single, lucrative species.

This situation represents the proverbial "gilded trap," a term that arose in resource management literature to describe instances where short-term economic gain and the illusion of long-term stability tempt resource users and policy makers to undertake actions that are fundamentally unstable and unsustainable.[5] And the lobster fishery continues to grow; the total numbers caught today are orders of magnitude higher than those landed a generation ago. For example, State of Maine records document that from 1950 to 1990 the annual Maine lobster landings ranged between 16 and 24 million pounds; by 2000 they were 57 million pounds, and by 2010 they had nearly doubled to 98 million pounds. A new record is set almost every year. Today, lobster accounts for more than 90 percent of the economic value of Maine's fisheries; the collapse of most other fisheries leaves lobster as one of the few remaining economically viable catches for much of the coast.

The contrast between lobster fisheries and most other forms of fishery governance is a striking illustration of the essential preconditions for success, and those that generate failure. Key among these is not just what the laws and institutional structures state, but the unstated and often unintended incentives to which the actors in the system respond. Understanding the commonalities of sustainable institutional design, and the pathologies of those that are not sustainable, is key to a science of open spaces that embraces social as well as biological means of maintaining large, functioning ecosystems and the communities and cultures that rely on them.

Emergent, Place-Based Governance

So what makes the lobster fishery an economic and ecological success amid a world of declining fish populations? To a certain extent, it may just be the luck of living in a period where ocean currents are favorable to the survivorship of young lobsters, guiding them from the open ocean, where they spend their early, planktonic life stages, to the rocky nearshore substrates, where they have access to food and shelter as adults. Another contributing factor has likely been the loss of cod and other predators, leading to increased survivorship of juveniles and young.[6]

From a governance perspective, the specific reasons for the relative success of the lobster fishery are likely threefold. First, the State of Maine requires a double gauge for measurement of harvestable size. Animals are excluded from the harvest for being too small or too large. This harvest limit serves the dual function of protecting potential breeding stock—that is, the younger lobsters—as well as the large older lobsters with the highest fecundity. In addition, females carrying eggs receive a *V* notch in the tail and are protected from harvest as long as the mark remains. Typically, this allows females to spawn and molt several more times before potential harvest. Keeping *V*-notched females or lobster that are too small ("shorts") is heavily censured within the fishing community, with social norms proving to be far more effective than law enforcement in sustaining local resources. This is a notable contrast to other fisheries, which are primarily regulated through external enforcement, and demonstrates how effective governance can set up the preconditions that promote self-regulation, which minimizes expensive and often inefficient external policing.

Second, lobster traps are intentionally inefficient, using a design that has remained largely unchanged since the 1800s. Though now made of light, durable wire rather than wood, which was widely used until the 1980s, a lobster trap is still essentially a box with a funnel at one or both ends. Modern traps now also have an outer chamber or "parlor" in addition to the main chamber where the bait is located. The funnel design allows the animal to find its way both in and out.[7] Data from video studies suggest that, in some instances, seven or more lobsters may visit a trap to feed for each one that is caught.[8] In this way, lobster fishing is a bit like ranching, with individuals fed and captured numerous times before they are harvested when they typically reach legal size at eight or more years of age. In this way lobstering supplementarily feeds many lobster, and the inefficient design conserves much of the resource. This creates a vested interest in the fishery; for example, some communities will as a group set their traps early in an effort to lure the "bugs" into their fishing community's territory. Once animals are there, the traps are inefficient enough that the lobster are not overharvested.

Finally, organization at the local level allows for stewardship that does not occur at larger scales in other fisheries. Relatively small, tight-knit lobstering communities are notable for their built-in incentives for conservation (which includes the aforementioned peer pressure and norms such as V notching females) and their effective communication via radio and talk on the dock, which results in the means and motive to self-regulate. Lobster fishermen own their boats, and so are not beholden to distant owners, but are very directly affected by the views of their family and neighbors. On some occasions fishermen who have not respected local tradition and have taken young lobster or otherwise violated the interests of the community have found themselves excluded from the fishery or their boats sunk. Furthermore, each harbor or community defends its own fishing territory.[9] Though lobster are mobile and capable of traveling great distances, especially as they move inshore during the warmer months to molt and mate (and offshore again in the fall as the water cools), the local nature of the fishery means that catch efforts are, to an extent, self-regulated by lobstermen on their respective fishing grounds.

In addition to the self-governance and stewardship of individual harbors and privately owned lobster boats, a number of state-designated fishing zones (fig. 3.3) are distributed up and down the coast that provide regional parameters for management that can be tailored to broader local conditions such as bottom type or coastal geography. Fishermen cannot transfer to other zones unless their move is permanent, and often they must apprentice by working as crew on a local boat for at least a season before being integrated into a new fishery. All of these factors create a vested interest on behalf of the community in locally based stewardship, which in turn represents what the researchers at the Santa Fe Institute called a complex adaptive system that couples social and ecological factors. The "agents" in the system respond to a simple and well-articulated set of "rules" that in turn lead to a complex, but predictable set of outcomes.

To demonstrate the role of this unspoken rule set in organizing and sustaining this connected social and ecological system, University of Maine economist Jim Wilson and colleagues used computer simulations of rule-based decision making by lobster fishermen to approximate their actual behavior on the water.[10] The model simulates the learning processes of multiple individuals, as well as competition among different

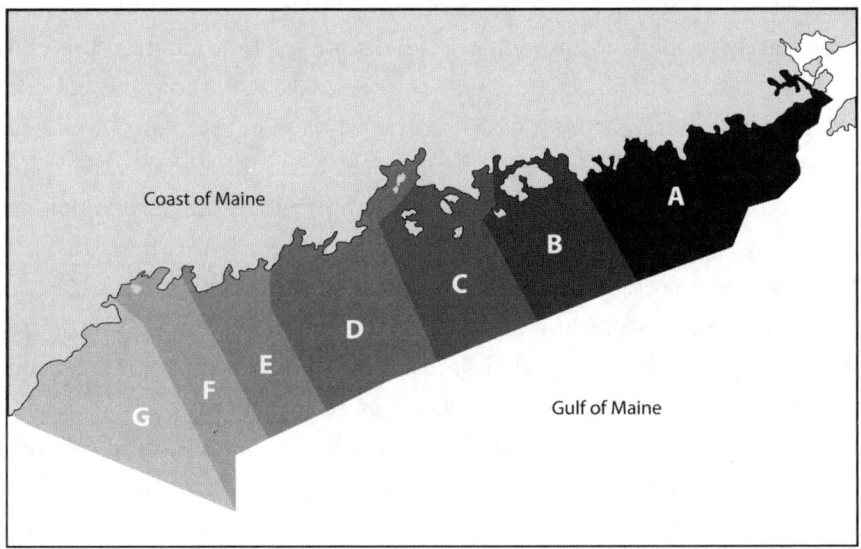

Figure 3.3. *Fishing zones distributed along the coast of Maine. Local "zone councils" set their own harvest regulations that are responsive to local conditions (as long as they are within the guidelines of the State of Maine), generating an adaptive framework in which policy is responsive to local conditions.*

groups of individuals, in a complex, changing environment, and contrasts the results with those stemming from the actions of forty-four actual fishermen. In doing so, the model analyzes fine-scale patterns emerging from a broad-scale, socioecological process.

In the model, the lobstermen set their traps, evaluate the performance, then respond to the lessons learned about their relative success in their patchy environment. The longer the simulated fishermen continue to search for better lobstering grounds, the more likely it is that they encounter one another. When they do, they gain an increased understanding of each other's fishing strategies; in this way learning is distributed naturally, which leads to collective action as both individually and communally the fishermen assess and learn. The results of the model, as with those from the actual fishery, suggest that collective action and sustainable resource use is more likely to occur in any environment when it is consistent with the self-interest of the parties involved.

However, the system is not without its drawbacks. It is generally conceded that too many traps are in the water during the peak summer harvest months. Some biologists advocate that based on the age of sexual maturity and potential capture, the minimum catch size should be raised moderately to allow young lobster more time for reproduction prior to harvest.[11] From a financial perspective, wholesalers still hold the bulk of the power in setting prices, which at the wholesale level have remained steady and even declined, while retail prices have, for the most part, increased. This drives even more consumption of the resource such that an increasing number of biologists and regulators worry for the industry's future and increasing efforts are being made to curb entry into the fishery. There remains the question of whether the capture of forage fish to use for lobster bait is harming other parts of the marine ecosystem. Yet even with these flaws considered, the lobster industry generally works in sustaining local economies and culture because it is grounded in the assumption that its resource is dynamic and unpredictable. In response, practices are both conservative and intrinsically adaptive, informed by signals (such as trap success) between the resource and the resource users, and among the resource users themselves. These differing signals generate the self-organization by which the internal social, ecological, and economic dynamics allow the fishery to be responsive to change with limited amounts of slow and politically costly external governance from state or federal entities.

Command-Control, Top-Down Governance

By comparison, fishing for cod (and other groundfish) is the antithesis of fishing for lobster, because it assumes the predictability of the species, as well as a large measure of anthropogenic control. Groundfish harvests are also managed at much larger scales.[12] These factors have all led to a markedly different set of ecological and social outcomes than those for lobstering. In recent decades, federal managers have viewed consolidation as a means of restricting catch effort, so rather than numerous small to midsized boats (35–80 feet) distributed along the Maine coast (fig. 3.4), there are now primarily a few large boats (80+ feet) located primarily in Massachusetts. As the boats left Maine and the smaller ports, many of the industries that supported them disappeared as well, which had cascading effects on local economies when everyone from seine net repairers to diesel mechanics lost their jobs or relocated to larger ports. These losses can, in turn, have dramatic effects on the rest of the local economy, leading to closure of local grocery stores and a general decline in necessary community resources.[13]

In addition to differences in gear, the scale of the groundfishing industry is entirely different from that of the lobster fishery, which has crucial implications for the sustainability of the fishery because it influences how vested people are in a single place. Trawling boats are relatively large and mobile compared to lobster boats. Although lobster fishing is confined to nearshore waters and discrete community fishing grounds, groundfish harvesting is typically restricted by time or catch effort (days at sea) or actual catch (various quota systems), but not by place.[14] These boats can roam anywhere and are not constrained by local interests, the need to conserve local resources, or community-based social norms or incentives. The combination of these factors often creates a "derby fishery," wherein boats race to capture the maximum amount allowed by quota, or inspires the "roving bandit" strategy, wherein patchy resources are selectively exploited before the boats move on to another area.[15] These methods encourage a race to the bottom, because few incentives exist to conserve or sustain resources. Several years ago, I asked the owner of several

Figure 3.4. *A midsize dragger (about 55 feet long) leaving port. These vessels when fishing for groundfish such as cod haul a net along the ocean floor. The elevated outriggers stabilize the boat while the net is being towed, and the drum at the rear of the boat is used to haul up or let out the net.*

Maine trawlers about her long-term goals. Her response? "To get filthy [expletive] rich." According to the perverse arithmetic created by the governance structure of the fishery, fishermen are encouraged to overfish. Even when only a fraction of the original number of fish remain, if a single boat can corner most of what remains, that vessel can come out ahead of others, and is compelled to do so. This competitive aspect is exacerbated by the fact that fishing permits themselves are immensely valuable; as the number of available permits declines, the value per permit rises so boats need to fish harder to make more money to cover rising costs. However, due to federal regulations, the fishing season is getting shorter. All of this is perceived by fishermen to lead to greater numbers of accidents and greater risk.

The rules favor large boats that can take fish from throughout the Gulf of Maine over diverse local fisheries composed of greater numbers of smaller family-owned operations.[16] The large scale of the groundfishery leads to fundamentally different management outcomes. Catch quotas exist primarily for those fish that are brought to port, with nontarget species captured at sea often thrown dead into the ocean as bycatch.[17] Multiple quotas can increase this problem by allowing fishermen to choose the fish with the highest market value that day and discard the others. The allowable bycatch can be as high as the legal catch of target species. Fishery models (i.e., computer simulations) attempt to make allowances for bycatch, but are inaccurate because bycatch is highly variable in space and time and is rarely reported. The deleterious impact of trawling on the health of fish stocks has continued to increase with improvements in technology, but the essential challenge remains: how to match the efficiency of the gear to the limitations of the ecology of the ecosystem.[18]

Evolution of Perverse Incentives: How Ecologically and Socially Destructive Policy Is Developed and Sustained

A review of the history of the groundfishery is useful not only for understanding contemporary problems in the fishing industry, but also for illustrating underlying pathologies in governance of natural resources in general. After World War II, early attempts at science-based ocean fishery management dictated a large-scale, single-species approach.[19] The science of management, then as now, tended to emphasize reductionist models of the most quantifiable elements, rarely asking whether these parameters were ecologically relevant.[20] This approach focused both management and science on area- and species-specific populations called "stocks." Assessments of stocks have remained to this day the cornerstone of fisheries management and are largely a reflection of the constraints of modeling, rather than those of fish or fishermen.

In 1950, an early attempt at large-scale management by the Inter-

national Commission for Northwest Atlantic Fisheries concentrated almost exclusively on commercial species within large "statistical areas"
that were thought to correspond with major geographic zones or fishing grounds (e.g., the Georges Bank, Grand Banks, Scotian Shelf, etc.).
This division of statistical areas and focus on stocks established a kind of
"intellectual path dependency" that has persisted to the present.[21] The approach was further codified in the 1976 Fishery Conservation and Management Act and subsequent legislation, with a myopic focus on relatively
undifferentiated stocks at large scales, setting in place an institutionalized
approach that all but ensured the many failures of ocean fishery management today.

The most significant implication of this intellectual inheritance has
been a scientific approach that simplifies complex ocean systems by treat-
ing individual species as if they are independent or isolated entities. The
core concept of the single-species theory is the assumption that the future
size of individual stocks is correlated with spawning stock biomass, which
in turn is determined by how much fishing occurs. This relationship was
initially chosen because it was considered clear and easy to measure. But
in reality little direct empirical evidence supports this correlation. The
lack of a firm theoretical or conceptual basis has thus resulted in an inability to predict future recruitment beyond applying an average value
from past assessments. Moreover, errors of measurement of past stocks
occur on the order of 30–50 percent, severely limiting the reliability of
even the few assessment tools that do exist for scientists and policy makers.[22] Fishermen themselves ridicule this obvious lack of reality, with the
following comment typical in a process marked by increased frustration
with the science and management:

By God those people [fishery managers] are stupid! Year after year
they come out here with their charts and graphs and measuring
tools and go to the same spot at the same time and try to catch fish
so they can compare this year's stock with last year's and 10 years
ago and so on. And when you tell them that's dim, that that's not
going to tell them anything, they mumble about "reliability" and
"sampling procedures," and like that. Jeeesus! Don't they understand that fish swim?[23]

A corollary to the stock-based approach is the assumption that ecological interactions among multiple species and the environment are minimal. This means that management decisions tend to be based on only a small number of easily measured variables. This then hampers adaptability in response to potential environmental change, because many of the most important interactions across trophic levels, or between the environment and fisheries, such as predator-prey interactions or competition/mutualism between species, are largely overlooked. This leads to limited capacity to understand the complex relationships involved and poor predictive ability of most fisheries models.[24] When most components of the system, and the interactions among them, are ignored, the synergistic and emergent outcomes of ecosystem interactions are missed, resulting in stock assessments essentially being a trailing indicator of fishery health, rather than the leading indicator they are meant to be. In this way the system is extremely effective at capturing dramatic change in the system—after the fact. This is a fundamental pathology contributing to poor management and collapse of ocean fisheries, for as we will see in later discussions of complexity and resilience, understanding cross-scale interactions is key to sustainable management.

Yet, even if the science were accurate in providing information about the system, what is ecologically sustainable is not always politically expedient. For many years, the shortcomings of the stocks-based approach served the interests of all parties by providing scientists with ease of measurement and fishermen with the relative flexibility to influence management and governance processes.

Although regional councils of local interests were established under the 1976 Fishery Conservation and Management Act to allow local input in management decisions (a seemingly egalitarian approach in allowing public access), in reality, a decision-making process involving primarily industry members instead meant that economics would often trump biology, to the long-term detriment of the fishery.[25] In a system where it is most essential to couple ecological and social processes, groundfisheries management did just the opposite by isolating the regulatory process from issues related to scientific uncertainty (fig. 3.5).

In a dynamic world, there will always be a lack of precise knowledge; rarely do we have the ability to predict the interactions among multiple

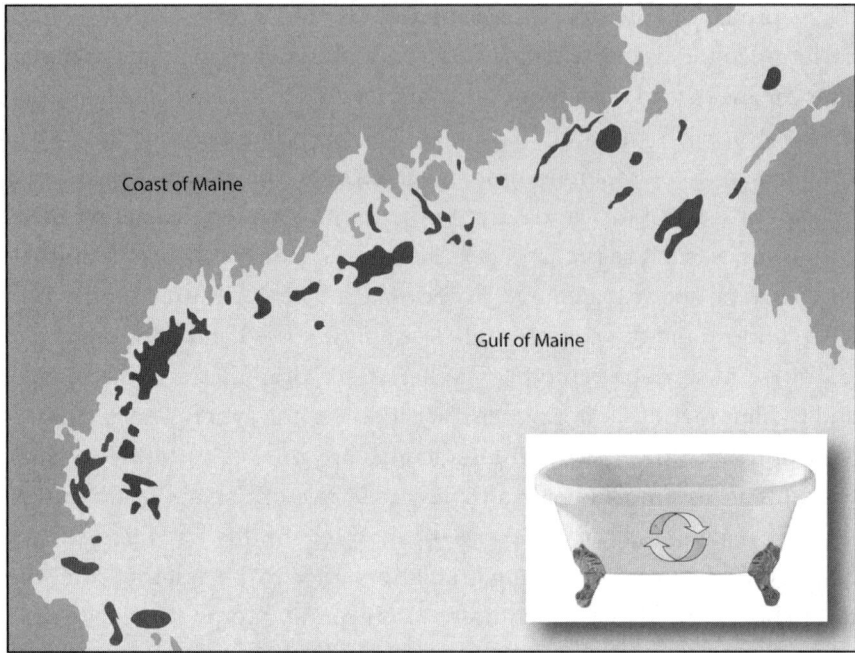

Figure 3.5. *Computer models typically used by fisheries managers view the Gulf of Maine as essentially a bathtub, an amorphous body of water where fish essentially move around at random. (Dark areas above represent historic spawning sites.) However, fishermen for more than a century have recognized a series of discrete spawning grounds and populations where overharvesting of a location does not result in the take of the fraction of the whole, but the elimination of the entire population of a single place. Evidence from Canada suggests it can take many decades at best (if ever) for these discrete populations to return once they have been eliminated (spawning ground image after Ames, 2004).*

variables. The challenges, therefore, are to design institutions that are durable in the face of uncertainty, to find ways to connect biological responses in natural ecosystems to human actions through policy, and to build sufficient slack into the system so that the inevitable mistakes are less costly. Even if accurate predictions of optimal harvest were possible, the ability to achieve that goal in practice would remain nearly impossible, because harvests undershoot and overshoot target levels through time. The groundfishery example illustrates a fundamental limitation in the way humans perceive and attempt to organize and manage the world, an assumption of predictability and control that is supported more by faith than evidence or reason.

Fate of the Cod Fishery: The Interplay of Ecology, Policy, and Technology

In the face of misguided incentives and simplistic assumptions in fishing governance, fish populations and communities the world over have undergone profound changes, which have accelerated in recent decades.[26] Fishery management thus represents a grand, if unintended, experiment in what happens when technology outstrips reason and common sense— a primer in the folly of coupling dynamic biological systems with rigid, top-down forms of governance. The consequences of removing critical links in the food chain through the overfishing of key species, and of ignoring important relationships among species, highlight the importance of conceptualizing fisheries as a complex adaptive system.

One of the clearest examples of such perturbation to the structure of an ecosystem is the decimation of cod in the northwestern Atlantic. Cod have not only been ecologically important as a top carnivore in the system, they have been economically and socially important in the northwest Atlantic region for thousands of years.[27] By the early seventeenth century, European powers had developed fishing stations on islands along the Maine coast for capturing fish and transporting them back to their respective nation-states, while early New England colonists had started to exploit inshore fisheries for their own subsistence. By the end of the seventeenth century, New Englanders began to fish offshore on the shallow continental shelves, or "banks," in the Gulf of Maine. Because the coastal waters of Penobscot Bay and eastern Maine were found to be highly productive, large numbers of small, inshore groundfishing vessels put to sea. By the early 1800s, nearly every coastal village was engaged in near-shore fisheries.[28] These fishing grounds would sustain local communities for nearly two centuries.

Groundfishing was originally conducted from sail-powered vessels (fig. 3.6) using hand lines, each of which carried a single hook. By the 1860s, this practice was replaced by tub trawling, which involved the use of long lines, each carrying as many as 1,800 hooks. Fishing with hooks,

Figure 3.6. *In the age of sail, fishing with hooks had considerably lower catch efficiency than modern technology, and although it undoubtedly affected fish populations, the take was not close to that of later, more mechanized gear that could target fish when they were in spawning aggregations and most vulnerable.*

even large numbers of them, was still relatively sustainable; when fish are in spawning aggregations and most vulnerable, they are also not eating and therefore are less vulnerable to capture by hook. This all changed with the advent of towing large nets to capture fish. The purse seine came into use in the 1870s, and in 1893, the first otter trawl (which uses rectangular "boards" to keep the net open) was introduced. These approaches were at the time confined to shallow waters with relatively smooth-bottomed seabeds or to open water.[29]

However, by the mid-twentieth century, technological advances and commercial demand began to surpass environmental constraints. "Rock hopper" gear (fig. 3.7), which features wheel-like devices on the lip of the net, allows nets to be pulled through rough and rocky habitats that had previously served as refugia for fish.[30]

By the 1950s, the Maine and Massachusetts fleets had depleted Maine's coastal groundfish stocks and eliminated fish from large numbers of coastal spawning grounds. Vessels large enough to handle open water

Figure 3.7. *Detail of the fishing gear aboard the vessel depicted in **figure 3.4** as the net is coming aboard. The round disks on the leading edge of the net, called rollers, allow the net to be dragged along the bottom, giving fishermen more access to fish and the fish less respite from fishing pressure.*

left the coastal shelf to fish the offshore banks, while owners of small and midsized boats left fishing altogether, or shifted to other species. By the late 1960s and early 1970s, groundfish catches had declined steeply, but technological advances continued. Sonar allowed fishermen to locate fish directly, while Loran and global positioning systems allowed them to precisely mark the location of catches.[31] As one fisherman recently remarked, "With modern technology we can now catch every last damn fish in the ocean. The challenge lies not in catching fish, but in having the restraint to let them live."

Development of Place-Based Responses to Conserving Fisheries on the Coast of Maine

A decade or two ago, the future of coastal conservation in Maine looked bleak. Most of the fish were gone and fishing communities were demoralized, sharing their bleak outlook with the ranching communities of the

Southwest before the advent of organizations such as the Malpai Borderlands Group (MBG). The National Marine Fisheries Service, and even the
State of Maine, seemed entrenched in a cycle of denial regarding the fate
of their resources. Although one cannot underestimate how difficult the
future of fishing and nearshore ecosystems remains to this day, there are,
nevertheless, signs of progress. Alewife and other sea-run fish are beginning to return, while dams are being removed on major rivers—actions
that can only help increase marine diversity. Area management of eastern
Maine to control overfishing through the creation of local incentives for
stewardship similar to those in the lobster fishery is becoming increasingly central to fisheries management. Furthermore, midwater trawling
and other, even more destructive approaches to fishing are being regulated out of nearshore waters, with a growing emphasis on the quality,
rather than the quantity, of fish caught as a focus of resource management. All of these efforts are critical steps toward achieving adaptive governance and stewardship as typified by the lobster fishery. However, as
area management takes its first steps, it also enters a time of great risk.
As rules and cultural norms are established that will either sustain or sink
the fishery, players who benefit from the current system (e.g., owners of
large boats in larger ports to the south) will challenge every aspect of the
burgeoning one that, although more sustainable, still represents a financial loss to some of the dominant players in the current system.[32]

However, at the same time this debate is occurring at the level of the
New England Fishery Management Council in the cities of urbanized
Massachusetts where fleets of trawlers remain, hundreds of miles away
in eastern Maine, small coastal communities are taking the future into
their own hands. On the coast, prominent examples of the many evolving
place-based organizations, include the Cobscook Bay Resource Center,
Penobscot East Resource Center, and the Midcoast Fishermens Association. Each organization implements its own distinct approach and seeks
its own solutions based on local conditions, but all are focused on designing and building sustainable fisheries for the long term.

The oldest example in Maine is arguably the Cobscook Bay Resource
Center, founded in the 1990s by Will Hopkins. The center undertakes
sustainable approaches to conservation based on meeting the needs of
the local people.[33] In addition to working to sustain the scallop fishery, the

center has built a community kitchen, which helps increase the diversity of products raised from the sea, while boosting the local economy and reducing pressure on current commercial fisheries. A critical part of the strategy is to break the cycle of donor dependence and establish a triple–bottom line approach that helps the local community and ecosystem, while also generating support for the resource center. Hopkins's efforts are typified by coalition building and attention to a single large ecosystem (Cobscook Bay). Although this organization is small, it is extremely focused and disciplined, containing a truly place-based approach that appears considerably more sustainable than the conventional nongovernmental organization (NGO) model of large facilities and dependence on donors and debt.

The challenge for all these organizations is to reduce dependency on donor dollars by making their programs (to the extent possible) economically self-sufficient. Relative freedom from issues of private property and challenges over how to assign property rights to conserve open commons gives western Atlantic fishermen, in many respects, more in common with African herdsmen than western U.S. ranchers, whose system is intricately bound by a formal system of property rights. The question remains as to whether some form of distributed governance can be established that will leverage the individual goals and aspirations of these separate organizations in a unified place-based network approach. One potential alternative being experimented with is collective impact.

Collective Impact: Seeking Synergy and More Effective Local and Regional Outcomes

One model that has recently emerged from social science researchers for generating self-organization and adaptive governance has been dubbed "collective impact."[34] Collective impact has been used to great effect in urban environments such as Cincinnati, but has yet to be widely applied in working landscapes, though as we will see, it uses many approaches already embraced in large landscape efforts.

Collective impact (fig. 3.8) operates on the premise that much of conservation and donor funding has fallen short of meeting its goals of significant societal transformation because decision and implementation programs are too fractured and diffuse, generating competition among players rather than collaboration in what is often a zero-sum game. Using collective impact, a number of smaller organizations work together under the leadership of an impartial third-party, or "backbone" organization, which provides unity and focus for problem solving. By cooperating in a network governance program, local organizations don't just get a larger slice of the funding pie, the size of the overall pie increases.

Through this approach, governance is also more explicitly linked to financial, as well as ecological, sustainability by creating regional incentives for support that more closely match the scale of the problem with the scale of the funding and in theory reduce competition and redundancy by promoting better coordination among NGOs. As in the MBG example where The Nature Conservancy helped steward the foundation of the group to facilitate the development of monitoring and science programs, collective impact done well should allow communities and local NGOs to maintain a focus on their mission, while generating synergies among groups with different geographic focuses and specialties.

The Malpai example highlights the utility of backbone organizations, for there really were two MBGs: one during the era with TNC coleadership, and one during the era without. In the early years, TNC essentially played the role of a collective impact–style backbone organization, and was extremely effective at guiding MBG's process. Once TNC stepped back from a coleadership role, the power vacuum left by its departure was almost immediately apparent, and the scientists moved rapidly from being full partners in the process to having a subsidiary role. As noted in the last chapter, there are limits to what one neighbor can tell another. MBG leadership—the ranchers themselves—needed to be much more nuanced in their approach and did not have the ability to just tell people to follow the group's principles, as TNC coleadership was able to do. The decline of peer-review-quality science as a guiding part of the MBG's approach may have been in large measure due to this fundamental shift in MBG's governance structure, from being guided in part by a dispassionate third party, to one in which the leadership was mired in the day-to-day pressures of local politics, as well as having to worry about running a ranch and making a living.

In helping organizations and disparate entities work together, collective impact works at a fundamental level, principally because although a

The Five Conditions of Collective Impact

Common Agenda	All participants have a shared vision for change including a common understanding of the problem and a joint approach to solving it through agreed upon actions.
Shared Measurement	Collecting data and measuring results consistently across all participants ensures efforts remain aligned and participants hold each other accountable.
Mutually Reinforcing Activities	Participant activities must be differentiated while still being coordinated through a mutually reinforcing plan of action.
Continuous Communication	Consistent and open communication is needed across the many players to build trust, assure mutual objectives, and create common motivation.
Backbone Support	Creating and managing collective impact requires a separate organizations(s) with staff and a specific set of skills to serve as the backbone for the entire initiative and coordinate participating organizations and agencies.

Figure 3.8. *The conditions for collective impact are also essential for collaborative science and conservation in general. There are crucial roles for science in every step of the process, from developing a joint vision of the system (such as the Malpai Borderlands Group did with their model of rangelands in* **figure 2.4***); to data collection, which is essential for testing assumptions and better understanding the dynamics of the system; to backbone functions, which requires integrating data collection and analysis and coordinating diverse facets of the conservation program. (After Hanleybrown et al., 2012.)*

local leader is very much embedded in the social context of the community, an outsider can take both the local and the 30,000-foot view, which is key for maintaining long-term continuity. In addition to mitigating the potential for internal conflict, there is no substitute for having an organization led by someone whose complete focus is on sustaining the overall long-term mission, whose first thought in the morning and last thought at night is how to move the process forward, and who is not burdened by the challenge of balancing one's own livelihood and local interpersonal relationships with the needs of the organization.

The irony is that for years, funders viewed the ability of community-based organizations to be locally led as a mark of success. Often, when outside institutions engage with local communities, they rapidly cede their power to local interests. Collaborative impact, and my own observations from the borderlands and elsewhere, suggest this may be the wrong approach. Sustaining successful collaboratives requires not just local engagement, but also strong external guidance provided by a backbone or coordinating organization, an "honest broker" whose sole mission is to objectively promote local interests from a much broader context. This approach is easily articulated, but very difficult to achieve in practice; the history of large NGOs suggests that too often such organizations are focused first and foremost on their own interests, using local people and projects as flexible funding strategies to acquire resources that support the group's overall infrastructure and mission. This is, in essence, a form of conservation imperialism that, while providing some local assistance for a time, for the most part just mines local programs like extractive resources until funding opportunities dry up or the public relations opportunities decline, prompting the NGO to move on to the next project. The challenge, therefore, is to build a governance structure that empowers locally while being managed regionally (fig 3.9). Whether this is truly possible, or sustainable, is an open question.

Another reason for taking a broader, more regional approach embodying forms of network governance is that in a world of globalized markets and metaconservation, local organizations are no longer pitted solely against one another. From an organizational perspective, funding is increasingly a region versus region competition in a worldwide contest to attract donor dollars. Meanwhile, funders today realize that to have lasting influence and most effectively leverage their resources, they need to pick a few specifically targeted areas and go deep, rather than scatter their support around a number of half-finished and unsustainable projects. Collective impact is a means of building the regional synergy necessary to compete in a global funding market. But it is also a way to leverage funding to maximize its effect by having groups coordinate their activities so they can each do what they do best, rather than constantly fight over diminishing amounts of resources.

However, collective impact offers nothing fundamentally new, for many of these approaches have been tried for years. Examples range from the mammoth Collaborative Sagebrush Initiative that seeks to sustain sage grouse habitat across a third of the western United States, to the African Conservation Centre's role in establishing the South Rift

Figure 3.9. *Trawl surveys in conjunction with local fishermen document the density and diversity of marine biota, but also provide important opportunities for communication between fishermen, scientists, and regulators. Such collaborative science is one way to build unified network approaches to large-scale conservation.*

Association of Land Owners (SORALO) in southern Kenya (discussed in chapter 1). However, collective impact provides a coherent framework by which to articulate lessons learned. Yet there is a certain danger in that organizations attempting to apply the larger principles of collective impact lose sight of the more fundamental lesson that these efforts are at their heart still about trust and integrity, for especially in collaborative conservation, the ends never justify the means. An error too often committed in developing collaborative, place-based approaches is that trust and relationships are considered expendable. The experience of recent efforts at collaborative impact in eastern Maine provides an example of the kinds of traps organizations and individuals fall into when they, even with the best of intentions, sacrifice principles of integrity for what they may honestly consider to be the greater good.

New Approach, Old Pathologies

One organization that has embraced the collective impact approach is the Manomet Center for Conservation Sciences, dually based in Manomet, Massachusetts, and Brunswick, Maine. One of Manomet's attempts to apply the collective impact paradigm is currently under way in eastern Maine through the development of the Downeast Fisheries Partnership program. Building on the experiences of restoring alewife and other anadromous fish, Manomet is working with conservation organizations and resource centers to develop a collective, regionwide fishery restoration program for eastern Maine. Still in its infancy, the initiative tests the ability to revive large-scale ecosystem processes, as well as to forge distributive, place-based governance among a wide array of institutions scattered across more than 150 miles of coast. On the face of it, the Downeast Fisheries Partnership offers a new run at an old problem of reviving decades of fisheries and environmental decline.

However, the irony of the collective action approach as undertaken in eastern Maine is that although the stated goals and approach are new to the region, many of the old pathologies remain. Rather than being a true partner, Manomet has instead applied a process that appears to be inclusive, but in reality emphasizes considerable amounts of top-down control. Manomet gained a toehold in the region through the ideas and efforts of local practitioners, and then fired them to insert their own staff when the program was under way. In the Downeast Fisheries Partnership example, the lack of an ethical foundation will likely come back to haunt the organization, because trust and integrity are a crucial part of the process. Key potential member organizations are already shying away from the partnership because the group violated their core principles of trust and fair dealing within a week of designating the principles as foundational tenets of the organization's approach. But perhaps more fundamental, by concentrating power around long-time fisheries conservation players, the same ideas and concepts that have already had limited effectiveness are simply rehashed in a new guise.

The example highlights a challenge to developing regional conservation on the coast of Maine and in many remote rural areas, where a relatively small and insular number of charismatic leaders have dominated the scene for decades. Numerous initiatives have had different names, but the underlying social dynamics remain almost the same. As a number of observers who have worked with community-based fisheries in Maine have noted, the constraints lie not just with the pathologies of fisheries management, but also with the idiosyncrasies of a small and insular group of conservation leaders.

There is a real concern that much will be lost when this senior group of practitioners retires or moves on, but additional progress may not be possible until a new group with fresh ideas and new approaches appears on the scene. It is a catch-22, for although the decades of experience are needed to sustain the process, so too is the need to evolve and change. The fisheries conservation programs in Maine, although making some progress, are still to a large extent trapped by their own characterization of the problem and their approach to fixing it. For collective impact to work (as it did at the outset of the Malpai example), it must bring something truly novel to the table, not simply rearrange the deck chairs of existing institutional or social structures. The above example illustrates that although governance must harness self-interest to be effective and sustainable, governance also cannot be a slave to self-interest and must be ethical, inclusive, and transparent. In a science of open spaces, to maintain large and complex systems, sustainability not only lies with designing effective science and governance, it also rests with managing social dynamics. This highlights that governance and design for conserving large landscapes (or seascapes) must not only be ethically sound, they must also be fundamentally adaptive to avoid the traps of existing practice.

Concluding Remarks

In the previous chapter, the Malpai borderlands example demonstrated the value of using science as a community-building tool to bring together diverse individuals and organizations in large-scale problem solving, but also the importance of institutional structures in sustaining science and the associated adaptive capacity. There is a fundamental disconnect in the nature of rewards, with relatively tangible short-term economic or social

Figure 3.10. Though ecologically extinct, cod are very much still a part of the culture of coastal Maine. Here in a "cod relay," teams wearing fishing gear race carrying dead cod down the main street of North Haven Island, Maine. The image shows the important role groundfishing continues to have in the culture of New England.

benefits frequently trumping more complex and nuanced long-term environmental concerns (fig. 3.10). A crucial role of governance and institutional design is to generate alignment between short- and long-term benefits by sustaining institutional processes of appropriate scope to address societal challenges.

In this chapter, we examined marine fisheries where governance has been more widely used than in rangelands. The open commons of marine systems serves as a proxy for open spaces on land. The federal groundfishery, contrasted with the state lobster fishery, illustrates how initial assumptions about scale and process can have dramatic repercussions for the long-term sustainability of resources, human communities, and large, open ecosystems. The most effective governance, emerging from a few simple rules and clear guiding principles, sets in place the adaptive dynamics necessary to sustain large, complex systems. New approaches, such as collective impact, currently test how to establish networked institutions that are securely grounded in local communities, yet are large enough to address complex social and ecological problems.

However, the Maine fisheries example also illustrates the pitfalls of existing governance arrangements. Across scales, from regional fishery councils advocated by the 1976 Fishery Conservation and Management Act, to local efforts to sustain coastal fisheries, the same approaches have over decades often failed because although with the best of intentions they seek to conserve diminishing resources, they inadvertently institutionalize existing pathologies. To paraphrase a statement attributed to Albert Einstein, one cannot solve problems with the same type of thinking that lead to them in the first place. This highlights the need for conservation and science systems that are adaptive and learn from their own and others' experience.

In the next chapter, we transition from deep immersion in two distinct systems (rangelands and fisheries) to stepping back and looking at the body of social and ecological knowledge that grounds adaptive approaches to large-scale science and policy. The integration of these perspectives provides crucial intellectual grounding for a science of open spaces, but highlights the need to take the process a step further toward additional synthesis of theory and practice.

CHAPTER FOUR

Conceptual Underpinnings for Preserving Open Spaces

Just as the constant increase of entropy is the basic law of the universe, so it is the basic law of life to be ever more highly structured and to struggle against entropy.

—*Václav Havel*

In the first three chapters we examined case studies from large-scale science and policy and witnessed how they succeed or fail based on their institutional frameworks. This chapter steps back from the application and synthesis of these examples to take a deeper look at the underlying principles, the foundations of which are physical (complexity and the second law of thermodynamics), biological (ecological processes and the role of scale), and social (the way people perceive, learn from, and address challenges). Understanding the conceptual underpinnings of large-scale, post-normal science and conservation is essential for developing overarching principles of practice that are key to conserving open spaces and broadly applicable across a range of systems.

Dynamic Foundations

All constructs, both physical and conceptual, are only as robust as their foundation. Like the foundation of a house, a conceptual foundation must be anchored to bedrock. Similarly, theory and practice should be grounded in what philosophers call first principles—basic or foundational propositions that cannot be deduced from others. Laws or principles that influence humans and their environment at a fundamental level are an essential component of any successful framework for understanding and sustaining large-scale conservation programs.

Successfully generating ecological and social resilience in large land-scapes and the human communities embedded within them requires conservationists to take into account the foundational laws that determine the intrinsic nature of complex systems. Theoreticians and practitioners approach systems differently, but the underlying principles are much the same. This was most poignantly driven home to me during early Malpai borderlands science planning sessions, when the group would often brainstorm ideas for sustaining their land. Around the same time, I co-chaired a workshop on complexity, thermodynamics, and ecology at the Santa Fe Institute, and it struck me that both the Malpai ranchers and the Santa Fe Institute theoreticians were grappling with the same question but from different perspectives: how to develop, sustain, and effectively guide self-organized systems. Ranchers don't use the jargon of academics, but their appreciation for the dynamics of complex systems can be just as sophisticated (and often more so) because they live them every day. They react instinctively to challenges and constraints, simultaneously juggling decisions based on fluctuating forces ranging from rainfall and cattle prices to habitat conservation and diesel repair. Maintaining their livelihood and way of life depends on understanding these complex interrelationships.

In the years since, I have observed this same phenomenon among fishermen in Maine and pastoralists in Africa, the Middle East, and elsewhere (fig. 4.1). What accounts for these similarities across continents and among different cultures? The reality is that we all exist in an energy-based world that obeys the laws of physics. The second law of thermodynamics establishes that energy organizes the living world and identifies the constraints to which all living entities must adapt, with dissipation of energy, or entropy, creating the hidden order that structures not just physical but also biological and social systems.[1]

Implications of Thermodynamics for Open Spaces

When we think about thermodynamics, we tend to do so through recollections of mind-numbing high school or college physics classes. However, when considered in the context of large, open spaces, what seems

Figure 4.1. *The exchange of information at the Maasai-Malpai meeting in 2004 highlighted the importance of linking social and ecological approaches to science and policy, and that the same principles are ubiquitous across continents and cultures.*

like dry and esoteric theory comes alive, as the dissipation of energy becomes a rich and varied lens through which to view the world.[2] The fundamental power of being grounded in thermodynamic principles is that it is scale-independent, working on levels ranging from individual organisms to our entire planet. As a common denominator thermodynamics is extremely useful in cross-walking between social and biological perspectives.

For example, an energetics/complexity-based framework, derived from first principles, demonstrates how poor fisheries governance considered in earlier chapters can unhinge the basic structure of entire ecosystems. In doing so, it illustrates the energetic implications of different approaches to governance.

According to thermodynamics, the ability of a species or assemblage of species to perform work (its *exergy*, or free energy) is proportional to its biomass and informational content (e.g., genetic diversity or overall

biological diversity).[3] For fish, such information includes the ability to locate historic spawning sites or feeding areas. However, there is more to
ecological function than simply the numerical quantity of fish. As noted
by theorists Jorgensen and Fath, "If there is more than one pathway away
from equilibrium, then it is the one yielding the most useful work (exergy) which ultimately moves the system furthest from thermodynamic
equilibrium that under prevailing conditions will tend to be selected"[4]
(fig. 4.2). This statement suggests how energy organizes social and ecological systems.

The same occurs in marine environments, where a dissipation of energy moves the ecosystem away from equilibrium in the face of dynamic
and ever-changing environmental conditions. Yet, as the example of ecosystem change in the Gulf of Maine illustrates, once the fishery passes
certain thresholds, not only does the ecosystem lose species or biomass,
but also the overall ecosystem itself is simplified and decoupled. The system loses the connections that held it together and maximized its uptake
of energy. The very fabric of the system unravels like stitching in a fine
tapestry.

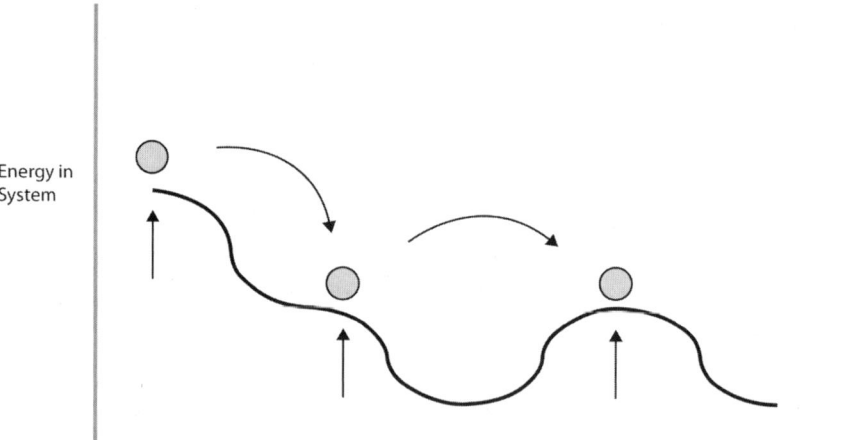

Energy in
System

Figure 4.2. *The complexity of social and ecological systems is a reflection of
the number and diversity of actors in the system and the energy inputs. With
increases in energy or complexity, systems are pushed farther from equilibrium.
Like geneticist Sewall Wright's concept of adaptive landscapes, in which fitness
is maximized at adaptive peaks (1932), the channeling of energy and resources
organize social and ecological systems and is a fundamental unifying principle
underlying a science of open spaces.*

Just as species interactions discussed in the previous chapter can cause nonlinear and exponential impacts, so too the impacts of the loss of species can amplify like shock waves throughout the system. Such a trend has been termed *devolution* by researchers examining long-term data from the Scotia Shelf off eastern Canada, where loss of cod has had cascading impacts on the rest of the ecosystem (fig. 4.3).[5] Just as in a recession or deflation in an economic system, a reduction of energy and resources reduces the overall ecosystem's productivity and ability to absorb other shocks or perturbations.

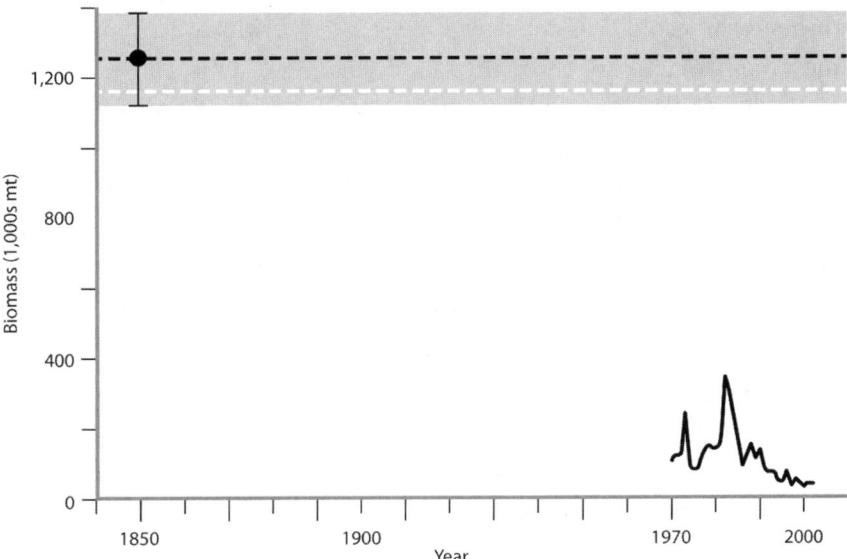

Figure 4.3. *In this figure, the shaded area represents historical biomass in the system based on fishing records from the 1800s, while the squiggly line on the right represents recent catch biomass. Though the amount of energy coming into the system is unlikely to have changed since the 1800s, the decoupling of the ecosystem and ecological extinction of a top predator, cod, forced a dramatic transformation of the ecology of the system to much lower levels of overall fish biomass. This is termed "shifting baselines." Though from the perspective of current fishermen the cod numbers may appear to peak in the late 1980s, they are still a fraction of historic levels. The graph depicts how fundamentally transformed the system has been over the last century, and how misleading short-term data can be without being viewed in the context of longer term information. (After Rosenberg et al., 2005.)*

In recent centuries, and especially in recent decades, fishing and other human activities have decimated marine predator populations and simplified food chains. For example, in the Northwest Atlantic, cod populations are estimated to be at 1 percent of historic levels.[6] Once the apex predator in the system, cod numbers are now so low that the species is considered *ecologically extinct.* The presence of fewer species reduces the thermodynamic potential of the system, with profound implications for the ecological function of the ocean and terrestrial ecological and social systems as well.[7]

One way to view the gradient of thermodynamic potential (or dissipative structures, as they are called from an energetics perspective) is through the ecological concept of *trophic levels,* which demonstrate the movement of energy through a food chain (fig. 4.4). These levels are highly interrelated, so a change in one level profoundly affects the others. One outcome is a *trophic cascade,* a top-down control mechanism in which, for example, a decline in a predator population results in the *release* of the population of its prey.[8] The prey population explodes, quickly outstripping its food sources, causing a decline in the prey population, and subsequently, in the predator numbers too. If these positive-feedback cycles continue, they eventually lead to the collapse of both populations.[9]

These interactions between predator and prey can have cascading effects throughout other parts of the ecosystem as the removal or addition of higher trophic levels in the food chain changes the distribution of resources in the environment. For example, recent data from Yellowstone National Park illustrate that when wolves were restored to the ecosystem, the large grazers became more wary, stimulating changes in their patterns of herbivory that were mirrored on the landscape by shifts in the distribution of resource patches. Aspen regeneration that had been nonexistent for decades soon occurred again, as elk were forced to become more mobile due to predation pressure that prevented them from grazing continuously in their preferred habitat, setting in motion significant changes to forest composition that may persist for centuries.[10]

Might the removal of cod and the subsequent vast increase in lobster numbers generate instability in the Gulf of Maine? We as a society are conducting that experiment right now. Archeological and historic data from fisheries provide an example of how changes in technology (e.g.,

Before   **After**

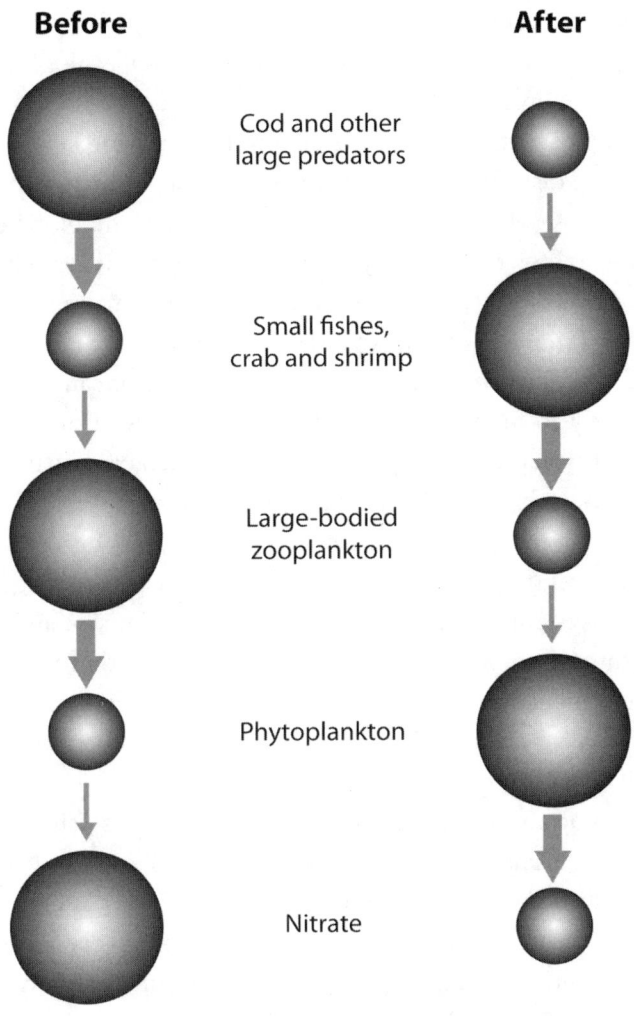

Cod and other
large predators

Small fishes,
crab and shrimp

Large-bodied
zooplankton

Phytoplankton

Nitrate

Trends in Ecology and Evolution,
Scheffer et al. 2005

Figure 4.4. *A depiction of changes in the tropic structure before and after heavy fishing pressure in the twentieth century, illustrating the cascading impacts of losing top predators in a system and how the western Atlantic has been transformed in recent decades. (After Scheffer et al., 2005.)*

human predation on marine resources) affect the trophic structure and energetics of coastal ecosystems.[11] Kelp forests are key ecological features of nearshore environments, crucial for supporting biological diversity and protecting coastlines from wave action. In observing kelp, ecologists recognize three distinct but overlapping phases of historic change.[12] The transitions among phases represent distinct shifts in the biotic potential of the system as food chains are shortened and key species are eliminated.

Phase one, lasting from four thousand years ago to the recent past, was a system dominated by apex predators such as cod and haddock. Archeological evidence from this time consists of significant numbers of bones, not only from large cod and haddock, but also from other large species—such as shark and sturgeon—that the Indians regularly harvested near shore. Invertebrates such as lobster are significantly absent from the record, suggesting that large predators had eliminated this lower trophic level.[13] Even during this aboriginal period, local peoples managed to fish down the ecosystem, as indicated by a decline in the size and diversity of species through time.[14] Archeological data suggest that the "pristine" system noted by colonial and European fishermen, with its diversity of fish and large numbers of cod and lobster,[15] was already an artifact of four thousand years of human action.

Phase two occurred during the 1970s and 1980s. As fish populations were exploited, fishing shifted to lower and lower trophic levels, until what remained were mostly mobile invertebrates such as crab and lobster. Freed from predation by fish, sea urchin underwent a population explosion in which they aggregated in feeding fronts that devoured and decimated kelp. The result was "urchin barrens" free of kelp except in turbulent shallow water zones. This loss of kelp significantly restructured the ecosystem by reducing habitat diversity and contributed to coastal erosion and lessened resilience of nearshore areas.

The third, or global, phase began in the late 1980s, as human action once again transformed the system in response to the creation of international markets for sushi.[16] In 1987, fishing began in earnest on the green sea urchin, and its populations were soon depleted in large regions of the coast of Maine. When urchin populations dropped below a crucial threshold, they could no longer control macroalgae and the associated kelp beds, resulting in a return of the kelp forests. However, without humans, cod, or other predators to limit their numbers, large crabs moved into the apex predator role. Higher and more diverse assemblages of algae, including nonnative bushy green algae, now occupy substantial substrate and may preclude urchin from returning to the ecosystem. The novel ecosystem that has emerged has a unique suite of players not seen before in the system and has developed a new rule set under which species, populations, and ecological processes are organized.[17]

This pattern of change through trophic cascades is evident throughout the Gulf of Maine, with a widespread increase in macroinvertebrates following declines in cod populations.[18] The Scotian Shelf off eastern Canada displayed a cascade in nearshore environments involving four trophic levels that had a pattern of change similar to that of kelp. The transition occurred from the mid-1980s through the early 1990s and was marked by a sharp decline in cod, which coincided with an increase in small pelagic fish and benthic macroinvertebrates such as snow crab and shrimp. Seals also appear to have benefited from the cod collapse, showing increased numbers in response to release of their forage base (small pelagic fish and invertebrates), while lobster populations have hit record catch levels in recent years.[19]

Whether these recent ecosystem changes are reversible is an open question. After World War II, catch records for cod and haddock in the North Atlantic suggest these fisheries briefly rebounded, apparently because of reduced fishing pressure in the preceding four years.[20] However, recent reductions in fishing pressure in parts of Canadian and eastern Maine waters have not yet had an impact. We have, in essence, transformed the system from one with vast numbers of midsize predators to one with vast numbers of detritivores. It is as if all of the lions, leopards, jackals, and other predators were removed from the Serengeti, along with all of the larger grazers, such as elephants and zebra, creating a landscape dominated by termites. In Maine, these "termites" (i.e., lobsters) are extremely valuable commercially, but the system is nonetheless diminished and destabilized,[21] its waters left with little of the immense ecological richness found only decades ago, while the economic richness is concentrated on a single species and a single group of fishermen.[22]

In fisheries across the globe, a similar pattern emerges.[23] As a system suffers repeated losses of apex predators, it experiences a downward spiral in function that leads to eventual collapse (fig. 4.5). Fishermen and fisheries biologists both note that following each bout of overfishing, the system rebounds, but to a lower level than before.[24]

Once its underlying resilience and capacity to adapt is eroded, the system goes through a cycle of decline in productivity and function. The same is true of other ecosystems. When rangelands lose significant grass cover or topsoil through overgrazing, they enter similar, positive feedback loops that promote further soil erosion, rather than retention (see Malpai borderlands model, figure 2.4). Like rebuilding cod populations, restoring rangelands is nearly impossible without the underlying abiotic and biotic resources such as topsoil and biological diversity to maximize energy uptake.

However, these declines in ecological function are rarely the result of only a single factor, such as overfishing or overgrazing. Our work in the borderlands suggests that a complex interaction between climate and grazing and myriad other factors typically leads to system change.[25] For example, a 1983 hurricane destroyed the seed resources of banner-tailed kangaroo rats, causing a collapse in their population. Because of the important role of their mounds as habitat for numerous other species, the loss of the kangaroo rats appeared to cause other species, ranging from burrowing owls to rattlesnakes, also to decline.[26]

Likewise in coastal Maine, fisherman and marine researcher Ted Ames discovered that viable cod populations appear to have persisted longest in areas where runs of alewife and other forage fish remained, suggesting that the collapse of cod was not due just to fishing pressure, but also to the loss of their prey base through dam construction and other human and natural factors.[27] There is also anecdotal evidence that eel grass beds were much more extensive in coastal waters before pollution, increased turbidity, and physical removal to clear harbors reduced their extent. These beds would have provided much more habitat and cover for small fish and lobster, supporting a diverse assemblage of species that, in turn, supported a more complex and resilient ecological system with apex predators. This provides another illustration of how complex interactions from loss of habitat and resources can transform ecosystems by reducing potential energetic pathways.

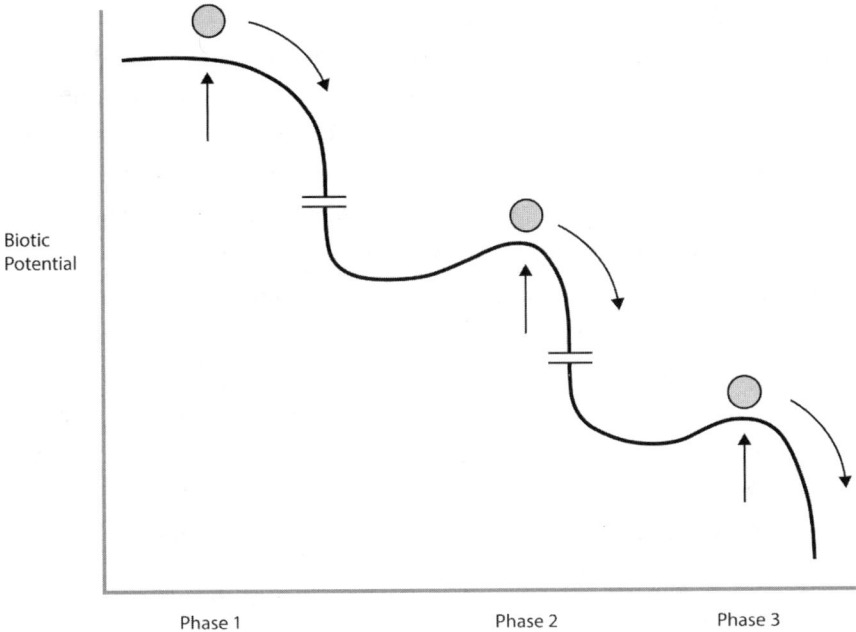

Biotic
Potential

Phase 1   Phase 2   Phase 3

Figure 4.5. *As the ecosystem is simplified and key elements are removed, the biotic potential declines. For example, in the western Atlantic, three phases of decline in nearshore systems lead to a reduction in the system's ability to take up energy. Once the ecological complexity is removed, it is frequently difficult or impossible for systems to rebound to previous levels, illustrating from an energetic perspective how loss of biodiversity can devastate ecological systems.*

Some fishermen contended that the return of forage fish through the reduction in midwater trawling is at least as important as managing the take of groundfish such as cod. In recent years, the reduction of midwater trawling in nearshore waters has provided something of a policy experiment with encouraging results. Work by Massachusetts Institute of Technology graduate student Sarah Hammit documented a return of forage fish and their predators inshore following a 2007 reduction in coastal trawling.[28] For the first time in decades, purse seines and other less invasive practices are once again in use, and nets and dories that have sat idle for years are now back in the water.[29] The return of forage fish to nearshore habitats is happening in time to recapture knowledge of minimally invasive fishing techniques such as small-scale purse seining, which allows for more selective harvesting of forage fish. This crucial information was all but lost as the generation that last used it neared retirement and the social connections among generations became more fragmented.

However, the above approach to fisheries as energetically based, complex systems is far from typical. It represents a transition in the way systems are perceived from static entities consisting of discrete populations that interact in predictable, linear ways to a more dynamic and adaptive approach (see discussion of groundfish management in chapter 3). Such a transition is not particular to fisheries or rangelands, but represents a much broader transition in science and policy in general. Understanding the meaning and root of such concepts as chaos and complexity that underlie a systems-based approach is crucial to appreciating the function of open spaces, and provides additional opportunities to explore the underlying theory that binds physical, biological, and social sciences.

Challenges to a Newtonian Worldview

Modern science in Europe began in the sixteenth and seventeenth centuries with the insights of Galileo, Copernicus, Newton, and others. This era of "Enlightenment" marked a return to the Greek commitment to "rational thought" following the focus in the Middle Ages on an almost entirely spiritual and phenomenological world. The movement toward rationality has influenced approaches to science and the humanities to the present and has generated perceptions of the universe as essentially a big machine, where if one understood and could account for all the parts, one would understand the whole.[30]

The principal challenge to the clockwork Newtonian worldview is embodied in the term *chaos theory*. Chaos has captured the popular imagination in recent years for its implications about the ability to predict the future, and its assertion that reality is, well, chaotic. But such is not really the case; chaos theory is nothing more than a collection of mathematical theorems demonstrating that the world is not linear. Chaos theory was popularized by a 1972 talk by climatologist Edward Lorenz titled "Does the flap of a butterfly's wings in Brazil set off a tornado in Texas?" The point of chaos is not that a butterfly can shift a vast weather front, but that at the very outset of a climatic event when the origins of the storm are but the slightest puff of wind somewhere in the Amazon basin, small changes in initial conditions can lead to vastly different outcomes.

This insight emerged from studies by Lorenz in the early 1960s. Through three simple equations with three variables, he sought to model weather on the comparatively simple computers of the day. However, the more detail that was added to the simulations, the less predictability occurred, because essentially random initial conditions, such as the flapping wings of a butterfly, resulted in weather that was not predictable beyond a few days, even though extensive knowledge of the mechanics of weather existed. The computer demonstrated that because of the profound complexity resulting from the idiosyncrasies of initial conditions, accurate weather forecasting was impossible. This lack of predictability in weather is essentially a metaphor for the lack of predictability in virtually all systems.

To understand chaos, it is useful to review its properties in two dimensions: space and time. An object whose behavior is chaotic in space is called *fractal*. The term was coined by Benoît Mandelbrot and was derived from the Latin to mean "broken" or "fractured." A general definition is a geometric figure that does not become simpler when you break it down into smaller parts. Nature is full of fractals; a braided stream or the vascular system of the human body is fractal, as are patterns of road systems, a fern leaf, snowflakes, the coastlines, dune formations, etc.[31]

Fractals are important for identifying discontinuities in the environment because most of the environment is self-similar across different scales. For example, when one sees a pronounced shift in patterns seen in aerial imagery, such as in a transition between forested and agricultural land, it indicates the underlying processes have in some way profoundly changed. Landscape ecologists use the *fractal dimension* of a system as a quantitative tool for identifying when thresholds are crossed in large systems.

Fractals also illustrate how scale is to a large extent an arbitrary construct, with the outcome of determining distance and space among objects frequently changing. In a classic paper on the topic of scale, Mandelbrot demonstrated how the length of the coast of England (fig. 4.6) actually changes based on the scale of measurement. An example I often use in my courses is the coast of the state of Maine. Viewed at the scale of a continental map, the coast of Maine is but a fraction of the eastern seaboard. However, viewed at the scale of miles, Maine's 3,500-mile-long rocky and braided coastline becomes longer than the entire East Coast, illustrating that the actual size of the environment may be very different from that depicted in Cartesian maps, and varies with scale so that not only do lengths of space seem different depending on mode of transport, they are different. So a bird or another highly mobile animal will, literally, travel a different distance to the same location than will an animal that experiences the landscape at a finer scale, such as a beetle or a mouse.

There is also chaos in the context of time, which is how we commonly view chaos. The outcome of time-chaos is the aforementioned sensitivity to initial conditions, as in the butterfly discussion above.

MIT physicist Michael Baranger used as an example a croissant. In the initial dough two points on the dough's surface may be right next to each other, yet as the croissant is rolled and folded to create the pastry the points move apart exponentially and predicting their end location based on initial conditions becomes all but impossible, even though we know precisely how the croissant was made. This facet of complex systems spelled the "death of reductionism,"[32] recognizing that simple linear relationships are rare in ecological and social systems, and rarer still where the two interact, which is the realm of most conservation and natural resource management. This destroys the illusion that humanity has absolute power to predict all things if only we can know enough of the details.

This insight has important implications for science and policy. Take the example of prescribed burns, for which such subtleties as small differences in wind direction, humidity, or soil moisture can lead to profoundly different fire behavior. The resulting forest composition caused by, say, a chance rainfall event that extinguishes a fire or a sudden wind that increases it, can exist for centuries, influencing nutrient flow, wildlife populations, timber values, and myriad other social and ecological factors.

Chaos is but one part of a much larger revolution in scientific thinking called complexity. A complex system is one in which numerous independent elements continuously interact and spontaneously organize and reorganize themselves into more elaborate structures over time—basically all living and most social systems. The original concept of complexity is

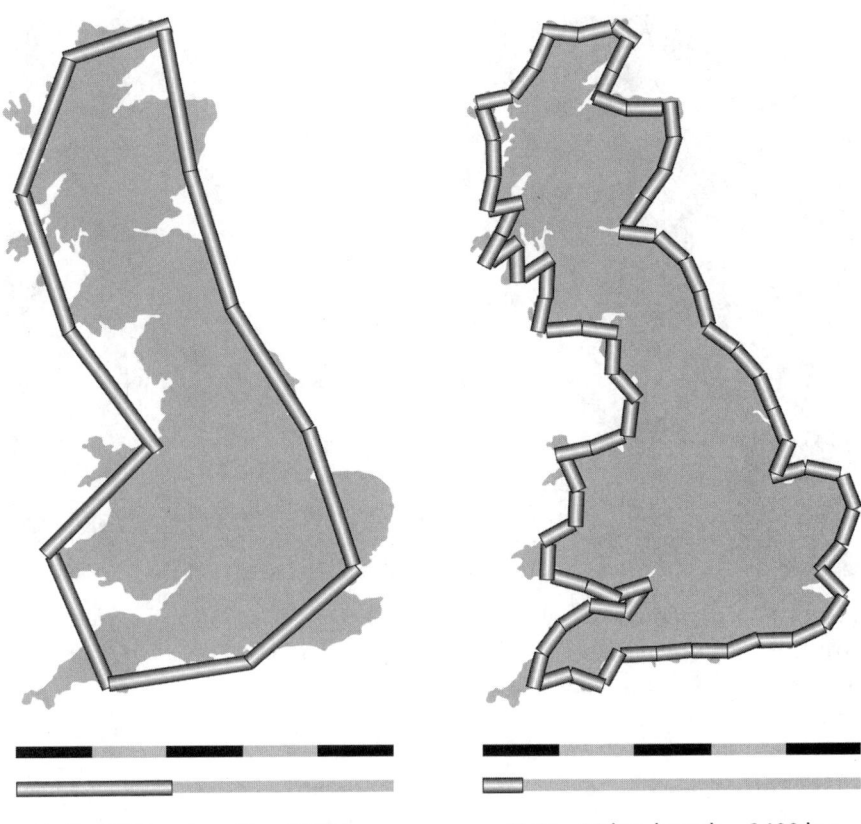

Figure 4.6. *The "coastline paradox," in which the smaller the increment of measurement, the longer the measured length becomes. Measuring a stretch of coastline with a 100-meter tape would yield a shorter result than would measuring the same stretch with a meter stick. The rougher the coast, the bigger this difference becomes, showing that distance and scale are often arbitrary, with distance changing with scale. (After Mandelbrot, 1967.)*

often attributed to a 1947 paper titled "Science and Complexity," by Warren Weaver. He posited that the complexity of a system equals the degree of difficulty in predicting the properties of said system even if its parts are known. Such systems can be said to have emergent properties when the whole is usually greater than the sum of the parts (fig. 4.7). This self-organization is a fundamental property of physical and biological systems, from flocks of birds to schools of fish. A few simple rules explain many of the complex and beautiful patterns found in nature.

Figure 4.7. Emergence occurs through a few simple rules leading to complex structures. For example, by orienting with neighbors, a school of fish or flock of birds emerges as a recurrent self-organized pattern in nature.

Complex adaptive systems (CAS) are a specialized kind of complex system composed of dynamic networks of many "agents," which may represent cells, species, individuals, firms, nations, etc., in constant interaction. Although the CAS paradigm comes from the work of the Santa Fe Institute, it can be traced back to older theories of cybernetics and general systems theory that flourished in the postwar era,[33] with some concepts dating back to the 1800s.[34]

Complexity and CAS paradigms are basically a collection of assumptions about the way the world works. A CAS behaves or evolves according to three key properties: order is emergent as opposed to predetermined, history is irreversible, and, because of the emergent properties, a system's future is often unpredictable.[35] The focus of this approach is on the collective actions of individual agents. Agents interact with their environment according to predetermined rules; some are foundational (such as the laws of thermodynamics), while others evolve (such as social interactions), but all contribute to the structure of biological and social systems.[36]

Both rangelands and fisheries contain examples of these complex interactions. In the previous chapter, the model of lobstermen's behavior in searching for lobster is an example of an agent-based, complexity-driven application in which a few simple rules have immense power in predicting the collective action.[37] Although we lack the ability to predict the precise actions of any given agent, the emergent interactions of all the agents reveal predictable patterns. The same holds in rangelands in the model of climatic interaction in mid-elevation landscapes that helped guide the Malpai science program by providing a framework for designing the points of focus (e.g., fig. 2.4). Under the model the dynamics of thousands of interactions across more than a million acres are distilled down to three core variables to capture the interaction of ranchers with their environment. The Malpai experimental research program was, in turn, designed to test the implications of the model at the level of large landscapes.

In the proceeding section, I have built on earlier case studies to show how a thermodynamically driven, complex worldview is critical for sustaining open systems—that the more "messy" the system, the more important it becomes to revert to overarching principles to bound the range of potential answers and outcomes. In working toward transdisciplinary synthesis, there is no choice but to revert to first principles to gain a common framework for addressing complex problems that span social and ecological perspectives. Understanding the physical foundation of action rules, successfully applying them to science and conservation, and embedding them in the design of governance is foundational to building a science of open spaces.

In the next section, we transition to looking at the foundations of ecology, a focus on the natural environment that is crucial for sustaining open spaces. But equally important, ecology is not just about the natural world, but also offers an understanding of scale and perspective essential to working across boundaries and for developing integrated approaches for sustaining large systems.

Ecological Foundations

The term *ecology* was coined by the German biologist and philosopher Ernst Haecke in the 1860s (*ökologie* in German). A combination of the root Greek words οἶκος ("house") and λογία ("study of"), ecology is defined as the study of the distribution and abundance of biological organisms, but it is increasingly becoming the study of interactions in all systems, including social as well as biological.

For the sake of simplicity, we can view the discipline of ecology from three discrete perspectives: botanical, zoological, and ecosystem. Each considers ecological patterns and processes from a different point of view, with implications for how communities, agencies, nongovernmental organizations, and researchers perceive and address environmental problems. The prime reason to review these foundations is that in conservation and management, the origins and context of ecological thought are often hopelessly muddled, with the outcome often being misunderstanding and misapplication of ideas. For example, the term *ecosystem science* is often used in federal agencies to refer to science where a bunch of variables are considered or a more holistic approach is taken. Although this is in part true, the reality is that the phrase ecosystem embodies a very particular approach to science and as a conceptual tool has very different implications for management than taking, for example, a population-based approach. So an understanding of the conceptual foundations of ecology is essential for effectively applying its principles, insights, and perspectives.

As we sort through the origins of ecology, it is important to recognize that science is not preordained truth, but the result of the perceptions, insights, and biases of real people, in real places. In the following abbreviated review of the discipline, key ideas are introduced via the stories of several select individuals who have been especially influential and who are representatives of a wider array of thought, recognizing that many important individuals and concepts cannot be included for the sake of brevity. We are hitting just the bare essentials.

The Botanical Perspective

Ecology as a science came of age in 1898 when Roscoe Pound and Frederic Clements placed a frame of a set size around vegetation, providing a systematic means for sampling the area within.[38] Doing so moved ecology from observational natural history to a quantifiable science through the introduction of the quadrat. With the development of this tool, the meter scale rapidly became a central part of ecological inquiry and the predetermined scale at which much of ecology would be practiced. As with the example of fractals in the previous section, in which change in scale shifted the relative size of coasts, this fine-scale approach also profoundly influenced ecologists' perceptions.

Although Roscoe Pound left ecology to become a Nebraska Supreme Court justice, Frederic Clements went on to become one of the seminal figures in the field. Raised in Nebraska in the closing days of the frontier, Clements witnessed the complete transformation of a natural system with the demise of the Plains Indians and the reduction of the great herds of American bison to heaps of sun-bleached bones across the prairie. These formative experiences, combined with deep religious convictions, led him to view humans as separate from the natural world and influenced his approach to ecology and the foundations of the discipline.[39]

In the early twentieth century Clements presided over an ecological dynasty at the University of Nebraska, where he crafted, through sheer force of personality and almost from whole cloth, a body of work that has had a lasting influence on ecology, conservation, and management through the present. Concepts such as vegetation succession and classification can be attributed to Clements's approach, which assumed linear and predictable pathways of environmental change. His students went on to assume leadership roles in the Franklin Delano Roosevelt administration, influencing federal responses to the Dust Bowl and other cases of environmental degradation through progressive policies that influenced conservation and environmental management for decades.[40]

In the 1950s the Clemensian paradigm of looking at plant communities as essentially a "super organism" with set community structure gave way to a more dynamic, individualistic approach in which components of the ecosystem differentially respond to the environment in a shifting mosaic of composition.[41] However, the framework of looking at vegetation associations as set entities remains to the present. Plant communities are still mapped and analyzed as discrete communities for the purposes of conservation and management.

The Zoological Perspective

In contrast to the botanical perspective that largely examined the interactions among groups of plants in fixed locations, the zoological sciences typically focused on populations of similar species and their movement or behavioral interactions.

Zoological ecology came of age in 1927 with the publication of the classic book *Animal Ecology* by Englishman Charles Elton, who introduced foundational concepts such as food chains and species niches. In 1933, American Aldo Leopold wrote *Game Management,* and the new field of wildlife biology was born. Leopold's work was a synthesis of academic and applied ecology stemming from his own experience and background in forestry, as well as interactions with seminal thinkers such as Elton. During this era, there was considerable interaction between the academic and applied branches of ecology. However, in postwar America, applied and theoretical factions parted ways. Applied ecologists maintained their focus on population dynamics and environmental variation, while more theoretical approaches increasingly embraced the emerging fields of genetics and evolution. In many respects, the greatest contribution to this new, more theory-laden perspective of the postwar era was the work of Canadian-born Robert MacArthur.

MacArthur grew up in Vermont, part of a remarkable family of artisans, musicians, and scientists. This diverse background contributed to the innovative approach that would make him one of the most influential ecologists of the twentieth century. After earning a master's in mathematics, he shifted his focus to ecology. This led him to Yale University to work with one of the seminal figures of twentieth-century ecology—G. Evelyn Hutchinson. MacArthur's combination of quantitative and observational skills would prove transformational to the discipline.[42] In the 1960s, MacArthur and compatriots forged a new synthesis of ecology by linking it with evolution. Through intense yet informal meetings at his cabin on South Pond in Marlboro, Vermont, MacArthur and his colleagues established the discipline of evolutionary ecology that would dominate the science of ecology for much of the next two decades. Perhaps most significant, MacArthur promoted a hypothetical-deductive approach by which researchers tested formal hypotheses through well-designed and replicated experiments. This rigorous and explicit testing of theory via experimentation improved the power of research, just as the introduction of the quadrat had decades earlier. MacArthur's intensive experimental studies made the science more robust and theoretical, but also often less directly relevant to the "real world." He recognized and accepted these limitations as necessary in a search for broad, unifying theory that would provide new foundations for ecology.[43]

Later in his career, MacArthur shifted away from the intense, reductionist science of the 1960s toward a broader and more synthetic perspective, reflected in the book *Geographical Ecology: Patterns in the Distribution of Species*.[44] His work in geographic ecology anticipated the broader and more integrative approaches such as macroecology that were to take hold in the 1990s and are still of great influence today.

The Ecosystem Perspective

The ecosystem approach to ecology is a significant departure from the other perspectives in that it focuses on the flows of materials and resources and energetics, rather than on individuals, communities, or processes such as evolution.[45] The approach has particular utility in large and complex systems where counting species is impossible; the simplification of the accounting process allows researchers to see larger patterns in complex systems. Ecosystem thinking was first linked to conservation and management through the writings of Aldo Leopold. Leopold was one of the first of his generation of American scientists to propose that humanity must see itself as part of the community of nature rather than apart from and above it, stating that "conservation is paved with good intentions which prove to be futile, or even dangerous, because they are devoid of critical understanding either of the land or of economic land-use."[46]

Ecosystem ecologists favor an integrated approach to understanding ecosystem organization, rather than emphasizing individual organisms. The ecosystem approach was expanded through the work of the Odum brothers, Eugene and Howard, and their colleagues in the 1950s.[47] The Odums were early pioneers of a human-oriented approach that was to characterize the subdiscipline in an era when most ecology largely ignored the role of humans.[48] Although Howard pioneered an explicit link between general systems theory and ecology using circuit diagrams to denote ecological interactions, Eugene's research broadened the application of the field in part by influencing the establishment of the first Earth Day in 1970, which he developed alongside legislative leaders such as Wisconsin's senator, Gaylord Nelson.

Clash of Perspectives

The concurrent rise of evolutionary and ecosystem ecology in the 1960s led to an intellectual battle: MacArthur viewed science as the art of abstraction,[49] whereas Ken Watt, Crawford S. (Buzz) Holling, and other system ecologists responded that to be relevant, ecology must address real-world problems. These debates over the relative merits of pure and applied ecology continue today; trade-offs between academic and pragmatic approaches remain points of contention. However, both perspectives agree that cutting-edge work must take chances. As Robert MacArthur observed, "There are worse sins for a scientist than to be wrong. One of them is to be trivial."[50] This approach is not rewarded in today's more calculated and incremental approach to science, a problem recognized by many leading ecologists, who are unified in their concern for the future of the discipline.[51]

Implications of Scale

As the example of fractals illustrated, the physical dimensions of an ecosystem are as important as its biological makeup. The species-area curve, perhaps the closest ecology comes to having a law, states that everything else being equal, as the size of an ecosystem increases, its diversity will also show a proportionate increase.[52] Ecologist Michael Rosenzweig called this the "tyranny of space," for ultimately the size of a habitat patch, or conservation area, directly corresponds to the potential diversity of species within it.[53] But there are innovative ways to capitalize on the constraints implied by this relationship, in essence gaming the system by establishing and promoting processes that maximize the ecological benefits of remaining land fragments. Much as the Malpai group promoted fire as a tool for facilitating ecological restoration at the vast scales necessary for sustaining ecological function, the core of conservation science—from the genetics of captive-rearing programs to reserve networks at the scale of the central Rockies—is essentially the process of figuring out how to make the natural world seem larger and more connected than it actually is.

The greatest threat to conservation and sustaining open spaces is habitat fragmentation. Like cutting a priceless Persian rug into eight pieces, what you get is not eight little Persian rugs, but remnants that are just frayed fragments of the interwoven whole.[54] The essential point of the science of open spaces is to craft a new synthesis among science, policy, and place, to support the processes necessary to sustain connectivity within natural and human systems. From the borderlands of the American Southwest to the arid rangelands of Kenya and the fisheries of the western Atlantic, the underlying issue driving these collaborative, place-based approaches is promoting the social integrity necessary to maintain fundamental ecological processes that sustain large intact and interconnected ecosystems.

However, the importance of size in a landscape is not just the outcome of linear relationships associated with area, but also reflects the emergent properties of scale. Examples from fire restoration to fishery management illustrate how addressing questions at larger scales results in fundamentally different outcomes (fig 4.8). Much of the art in conservation comes from determining the appropriate level at which to intervene in a system. Natural processes do not exist at fixed scales; rather, observers of a system select a level of inquiry implicitly or explicitly based on their own biases and background and sometimes even their desire for a particular outcome.

One method of assessing the appropriate scale of action is hierarchy theory,[55] which arose out of general systems theory, starting with the work of influential social scientist and systems thinker Herbert Simon beginning in the 1950s.[56] According to ecologist Robert O'Neill and colleagues in a 1986 Princeton series monograph: "All complex systems, including ecosystems, appear to be hierarchically structured as a natural consequence of evolutionary processes operating on thermodynamically open, dissipative systems."[57] As such, processes are constrained by larger scales but informed by, or composed of, smaller ones. Scale is largely contextual; the properties of surrounding levels have significant implications for understanding the level of interest. Therefore, one of the most

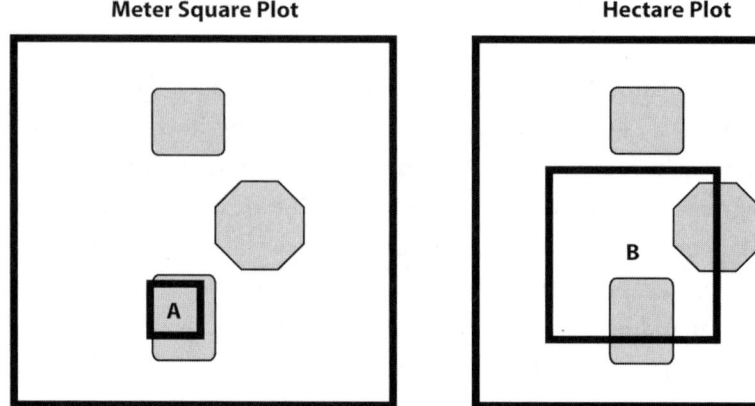

Figure 4.8. *The scale of an investigation strongly influences its outcome. In the example above, a meter plot (not to scale) in a single burned or grazed patch may indicate damage to the ecosystem. At the larger scale of a hectare, incorporating a range of burn or grazing intensities into the measurement documents a contribution of disturbance to diversity and ecological function. The emergent properties of these disturbances are exhibited only at the large scales at which these disturbances function. At smaller scales the measured outcomes are incomplete at best, and often extremely misleading.*

important considerations for conservation or policy design is the flow of knowledge and materials across levels in a hierarchical system. Lower levels can reveal a considerable amount of information about the properties of upper levels of which they are a subset, whereas upper levels frequently do not inform lower levels, but rather constrain or control them.

Hierarchy theory's applicability to ecology and natural resource management came into prominence in the 1980s with the work of Welsh ecologist Tim Allen and colleagues.[58] Allen was introduced to landscape and community approaches to ecology through his love of rock climbing. He studied "landscapes" of lichen on cliffs for his graduate work and soon realized that scale and perspective were entirely different concepts. Students are often taught that ecological organization exists in a hierarchy based on size. Populations are composed of individuals of the same species, communities are composed of populations of different species, landscapes are composed of different communities, and so on. However, Allen forged a perspective on scale that demonstrated that each of these ecological groupings was *scale-independent* and not defined by size. By this

I mean that populations are not necessarily smaller than communities, communities are not necessarily smaller than landscapes, and so on. One of Allen's favorite examples was of groves of giant sequoia that grow in "fairy rings," circles of adult trees produced by a parent tree that grew in the center perhaps thousands of years earlier. Thus, the spatial patterns and population dynamics of a single stand of redwoods transcend temporal variation in even such long-term processes as climate. Allen stated to students in his ecology classes, "Good and bad millennia come and go, but these trees survive."

This example illustrates how systems can be viewed from a range of scales or perspectives, with each level revealing very different structures, processes, and insights because of how organization changes across scales and contexts. As with twisting a kaleidoscope, the observer can adjust the scale or change the perspective to attain a different image. Similarly, the observer can change the scale and context to gain a different perspective to attain the most powerful insight or to determine the most effective approach to a particular situation or set of questions.

Therefore, an important distinction must be made between scale and type. Often, large spaces are viewed by default as ecosystems or landscapes, small ones as individuals or populations. However, a hierarchical lens adds a vast amount of richness to science and policy options by recognizing that constructs such as academic disciplines are not intrinsically linked to particular scales, but are instead distinct points of view—discrete ways of viewing the world.[59] The example above of ecological foundations stresses different facets of ecology as the outcome of perspective and not scale, with each subdiscipline viewing the world through a very different conceptual lens. This insight is especially significant to consider if we are to preserve open spaces, because, as we have seen in examples from rangelands and fisheries, success or failure invariably hinges on issues of scale and the associated approaches to problem solving. The greater the diversity of perspectives available, the wider the range of potential solutions to conservation and management challenges.

In summary, scale and perspective are key conceptual tools for developing more effective conservation of open spaces. When we link the scale of the ecology to that of the local social system, we can generate more emergent approaches to conservation and management that are better able to respond to social or environmental variability,[60] as the contrast between emergent approaches to governance such as the lobster fishery and the more rigid and prescriptive groundfishery demonstrated.[61] Applying these different approaches to science and policy, in turn, requires understanding that all constraints and opportunities are a function of the human mind's ability to see the nature of problems and generate creative solutions. Therefore, to craft effective solutions, it is critical to understand not just the range of ecological options, but also how people and institutions learn, process, and apply information.

Learning, Cognition, and Innovation

A science of open spaces is predicated on individual or organizational capacity for learning. Without learning, sustaining relevant science and policy at large scales within complex systems is impossible.

So what is "learning"? Learning emerges from the interaction of individuals and organizations and the social and ecological systems within which they are embedded. People within these entities learn from their interactions with each other and their natural and social context, but they seldom do so deliberately.[62] In the Malpai and fisheries examples, the institutional designs were not explicitly intended to promote learning (at least not for the local ranchers or fishermen in the system), but as a means to address threats or achieve positive outcomes. Learning is a self-organized process in which the more information is available, the greater the potential for success.[63] But these choices are constrained by the individuals, organisms, and institutions involved, their cognitive abilities, and their ability to seek the most effective choices from the total range of options.

Cognition as an Emergent Process

To understand how complex socioecological systems work, one must recognize that people are the primary catalysts for science, conservation, and management, and as such their underlying cognitive processes are as important as physical or ecological factors in shaping conservation outcomes.

The Santiago theory of cognition, first proposed by Gregory Bateson and further developed by Chilean researchers Humberto Maturana and Francisco Varela, states that cognition is the very process of life, including perception, emotion, and behavior, challenging the Cartesian division of mind and matter, with the mind not a thing, but a process.[64] Maturana and Varela offered a theory of cognition in which knowledge and meaning are understood from a biological and evolutionary standpoint. The brain is not so important in and of itself as is the network. Each cell is not important; they just hold sodium-calcium ions. The flow of ions constitutes the process that we call "thinking." In essence, the mind is the specific structure through which cognitive processes operate.[65]

Memory is the manner in which experience is encoded in the brain, and human action emerges from conscious and unconscious thought. Our brains contain an abundance of stored patterns that results from past experience; how we respond to a particular situation is largely determined by which neural pathway is activated.[66] These priming effects are key to unconscious selection processes, with all sensory inputs, including verbal and nonverbal human communication, activating particular patterns of neural connections. This is important for undertaking science and conservation, for it means we become almost hardwired to respond in particular ways to new circumstances based on past experience. Focusing events such as stressful or challenging experiences are retained more effectively because stress hormones activated by emotional situations enhance long-term memory.[67]

The periods of social and ecological change that precipitated science and conservation, as exhibited in the case studies, are consistent with the biology of the mind, illustrating the physiological reasons why periods of stress and conflict are also periods of opportunity. By contrast, in nonstressful situations repeated exposure over time is needed for new ideas to be "learned" and consolidated into long-term memory.[68] Culture is an emergent process stemming from the "accumulation of partial solutions to frequently encountered problems."[69] The human mind, then, is essentially a complex adaptive system with hierarchical organization. Properties observed at one functional level, such as memories or thoughts, are emergent from interactions arising from dynamic processes among lower-level entities, such as neurons that are the building blocks of cognitive architecture.

Cognition emerges as a consequence of continuous interaction between the system and its environment, which is why physical, ecological, and social systems, though considered in isolation, are for all practical purposes entwined. The continuous interaction triggers bilateral perturbations—problems or challenges to the system. In response, the system uses its functional differentiation procedures to come up with a solution if it does not have one already in its memory. This approach to cognition has been an important influence on complexity theory, especially in regard to issues or questions of epistemology, learning, knowledge, causality, and emergence. The Santiago theory of cognition integrates phylogeny and ontogeny.[70] With this approach to knowledge, the knower essentially becomes a complex agent within a self-maintaining framework.[71] This foundation is important because it provides building blocks for understanding the context within which individuals, institutions, and organizations exist and make decisions, reflecting a complex systems approach in which a few simple rules lead to the emergent processes and interactions by which decision making, collective action, and learning emerge.

In summary, from a cognitive perspective, effective decision making provides a theoretical foundation for three fundamental premises. First, that successful transformations toward effective science and policy at large scales tend to emerge from informal networks that help facilitate information flows, identify knowledge gaps, and create nodes of expertise. Second, joint understanding of processes in commonly held conceptual or mechanistic models is essential for developing the common ground necessary to generate effective action in response to challenges as they emerge. Third, collaborative approaches are essential for building the social capital needed to sustainably address challenges in large and complex systems. These perceptions of the outcomes of cognition mean one must consider thought not just at the individual level, but also at the collective level, where action occurs at large scales.

Distributed Cognition

The mind is largely a product of its environment and the collective understanding of those around us, with memory and perception both an individual and a group process.[72] Organizational theory early on recognized collective understanding as an adaptive process.[73] Studies of juries and scientific thought have explored the intersection of individual and collective thought. From these studies two points emerge. The first is that we are highly influenced by our environment. The second point is the role of collective action and interaction. Considering group versus individual thought may be much like contrasting the actions of a single ant with those of the colony. Although there may be some individual thought or action, there is also collective, distributive thought that is emergent, arising out of the collective action and understanding of the entire group.

Within the distributive cognition context, Keene State University biologist Tim Allen (not be confused with Tim Allen at the University of Wisconsin mentioned earlier in this chapter) studied the mental maps of lobster fishermen in Maine and Australia. Allen viewed the perception of the environment as both a collective and an individual experience. Individual lobstermen experience the environment through their own senses (such as clearing mud or gravel off a lobster trap when it is raised), or indirectly through equipment such as sonar. This understanding is shared with other fishermen through constant radio chatter and discussions dockside, or is passed on to future generations, such that a collective understanding emerges. Underwater features are named and understood every bit as well as those of landscapes encountered by farmers or ranchers— ironically, perhaps better, because in terrestrial landscapes people tend to stick to roads or well-worn paths, so they are repeatedly traversing the same environment and they often never see vast off-road areas. In open commons, such as oceans, everyone has equal access at least to experience the environment through sonar or other devices, so the collective ability to understand topography can be greater (shoals or other dangerous waters excluded).[74]

The point is that cognition is individual and collective, and this broad understanding forms the template through which we experience the world and understand and address challenges. Just as the collective understanding of the fishermen helps them both find fish and reduce costs and hazards, so too, distributed cognition is important in guiding collective action to generate sustainable approaches to environmental challenges. Therefore, although collaborative approaches to conservation are relatively time-consuming and expensive, the cognitive elements of theory and practice demonstrate that they are necessary. There are no shortcuts to durable design and reflective practice, for sustainable outcomes require cross-scale integration and development of networks between diverse individuals and organizations. The power of these approaches lies in collaborative frameworks that build on an understanding of individual and group cognition, and in developing institutions that take an intrinsically long-term view, matching the scale and resolution of the problem with that of the solution.

Cognitive Bias

Perception is not reality; our background, age, gender, culture, and innumerable other factors shape the way we perceive the world. In environmental conflicts, it is often not the facts, but their interpretation, that forms barriers to durable policy and effective science and decision making. Four kinds of cognitive bias influence not just individuals, but also the way organizations address problems. First, bias can occur from an incorrect inference between cause and effect. Second, biases can result from different perceptions of evidence. Third, bias can be induced by social interactions. Finally, biases can affect organizations, rather then individuals, arising from institutional design and goals. All four kinds of bias result in skewed reasoning that can interfere with attaining logical or optimal results; therefore, effective conservation needs to take into account and, where possible, mitigate issues of perception.

Cognitive bias is an intrinsic property of all human information processing. At the individual and collective level, dealing with complexity requires the ability to filter vast amounts of information and assemble it into some sort of predictive model. Thus, bias is a direct outcome of different individual and cultural perceptions, as well as the sorting process we intrinsically need to function in a complex and dynamic world. It cannot be eliminated at the individual level, though institutional design can be used to mitigate bias, as in the borderlands and fisheries examples, where the social frameworks (such as the Malpai Borderlands Group expecting openness and transparency) resulted in fundamentally different social interactions and science and policy outcomes. But the more the process challenges basic assumptions, the greater the time and investment that is needed to build common ground and effective process. There is also a systematic tendency to overestimate the likelihood that plans will succeed and a tendency to underestimate the potential for failure.[75] As resilience approaches discussed in the next chapter will illustrate, surprise is an intrinsic part of all endeavors, especially when working at large scales, and yet few institutional designs cope well with change or can adapt to or learn from failure.

As the complexity discussion and the case studies illustrate, the success of science and policy often rests with distilling a system down to its core components to attain a more subjective overview of the interconnections within the whole. Yet as an understanding of the function of the brain reveals, 98 percent of cognition is unconscious, so we are largely unaware of the process of thought and only experience the outcome.[76] This has significant implications for conservation and management, because the unconscious processing by the brain defines a potential set of solutions. As in the case of ocean fisheries, decision makers often assume they know the problem before they design a solution, seldom examining the underlying assumptions or seeking alternative approaches.

Reflective practice then requires conscious cognitive approaches that entail understanding one's own bias, and creating a system to test assumptions, adapt, or adjust to change: essentially the fundamental role of science. This was illustrated in the borderlands, where local knowledge, monitoring, and experimental research were all employed to understand the roles of climate, fire, and grazing and to test the group's premise that ranching was compatible with large-scale conservation. However, there are still considerable limitations to developing effective approaches to science and conservation, especially in large and complex arenas. The Malpai example also illustrates the challenges of maintaining relevant science in large systems.

One approach for understanding how perception limits action is the concept of bounded rationality. In the classic work on the topic, Simon[77] stated,

> Actual behavior falls short, in at least three ways, of objective rationality . . . :
>
> 1. Rationality requires a complete knowledge and anticipation of the consequences that will follow on each choice. In fact, knowledge of consequences is always fragmentary.

2. Since these consequences lie in the future, imagination must supply the lack of experienced feeling in attaching value to them. But values can only be imperfectly anticipated.

3. Rationality requires a choice among all possible alternative behaviors. In actual behavior, only a very few of all these possible alternatives ever come to mind.

Simon thus proposed that humans are *not* rational in the idealized sense of considering all the options and selecting the best action. Rather, people employ *bounded* rationality, in which decisions are the outcome of a limited set of alternatives based on their real and perceived constraints. From this, three points emerge: First, nearly all decisions aim at satisfactory, rather then optimum, outcomes. In most situations individuals or institutions are unable to comprehend or identify the optimum choice based on their own cognitive limits or imperfect information. This is a crucial insight into a key role of science, as in the large landscape studies in the borderlands: to reframe the question and enlarge the range of acceptable options.

Second, problem solving is situational. Not only do people choose among a selection of options, they also choose among a selection of problems, often deciding to put off one question or goal at least in the near term, to achieve a more desirable outcome in the long run.[78] The groundfishery example in the previous chapter illustrates this principle at work; fishery management is based on ease of measurement more than ecological relevance, with significant negative consequences for the environment and the long-term viability of the fishing industry.

Third, rapidly changing environments pose especially difficult situations because the very bounds of the decision-making space may be in flux. Bounded decision making requires that it conform to an established framework. In adopting a dynamic or adaptive approach, especially for policy makers or managers in a federal context, governance is largely about reducing uncertainty through addressing challenges within established zones of acceptance. This gives the policy continuity and the actors and their actions accountability. But it is also severely limiting.

An example of the indirect effect of implicit policy outcomes was a U.S. Forest Service employee who remarked at a Malpai meeting in regard to the challenges of prescribed fire, "I can burn a 100 times and receive little interest or recognition, yet if a burn goes wrong once, it is my career."[79] In this case of a highly risk-averse system, the preservation of long-term ecosystem function goals is eclipsed by fear of making a mistake. The federal system is not well equipped to address issues of long-term ecosystem health, because they fall outside of the bounds of localized decision making, yet the system is eminently qualified to assign and address fault. Under this decision-making framework, even though doing nothing would lead to certain decline of the ecosystem, it is preferable to the short-term uncertainties of taking tangible action.[80] In this case there was the courage and foresight to step outside the narrow confines of the agency to see the wider importance of taking proactive action, with external support from the partnership with the Malpai Borderlands Group providing the employee cover for his actions and enlarging the range of options.

Bounded rationality is procedurally rational; it has nothing to do with the overall logic of the situation. It provides a general formula for dealing with complex tasks within complex environments and highlights the importance of anticipating the range of solutions and building them into the overall policy and governance process.[81] Cultural artifacts and custom, as well as accidents of history, often play a significant role in determining the bounds of acceptable action. For example, in Western societies, perceptions of the role of women for decades influenced their ability to enter the workforce, and these constraints influence the bounds of rationality to this day. Often surprise events or social crisis cause dramatic change to these paradigms. The advent of World War II and the lack of labor facilitated women's movement into the workforce. Once this change occurred, new standards of social conduct emerged because of the path dependency of the system. This led to dramatic changes in women's roles in society.

Theories of Learning and Practice

The proceeding pages highlight the function of the mind, the individual, and the interplay with collective thought, as well as the process of applying information and learning within complex adaptive systems. One starting point for appreciating the implications of these interactions is the classic work on organizational theory by Argyris and Schön, of Harvard University and the Massachusetts Institute of Technology, in which people are thought to hold in their mind mental maps of how they will respond to a given situation. Few people are aware of the maps they use and the way they affect how they plan, implement, or review their actions. In terms of organizational learning, then, an institution is not a static entity, but at its very root a "cognitive enterprise," in which individual and collective thought processes are redirected toward collective goals.[82]

Organizations are therefore as much a reflection of human inquiry as the emergent process of system organization. This is key, for it implies that the way conservation organizations are organized profoundly influences their effectiveness. This may be obvious, but the institutional design of organizations is rarely considered when undertaking science and policy, and one frequently sees institutional designs that may be procedurally rational, while also containing elements that prevent the group from ever reaching its long-term goals. The role of a strategic process that openly considers pathologies in organizational design in circumventing barriers to progress is a crucial part of a science of open spaces to which we will return in the closing chapter. As beautifully illustrated in the Malpai example, relatively subtle differences in organizational structures either promoted effective action or prevented it. Movement between successful and counterproductive processes is often fluid and variable through time, with the outcomes often difficult to anticipate without careful consideration of the long-term indirect effects of policy choices.

In characterizing different approaches to learning in organizations, Argyris and Schön[83] were also among the first to propose a typology for understanding how different approaches to learning and institutional or organizational design affect an individuals' or organizations' effectiveness. In single-loop learning (fig. 4.9), the individual, group, or organization essentially modifies its actions according to differences between obtained and expected outcomes. The outcome of the process is that the decision maker does not seek the best outcome among competing objectives, but instead a satisfactory result for each goal, viewing the goals in isolation and to be addressed one at a time. Large ecosystem projects derived from a political process often exhibit these single-loop or "cybernetic"

Learning Loops

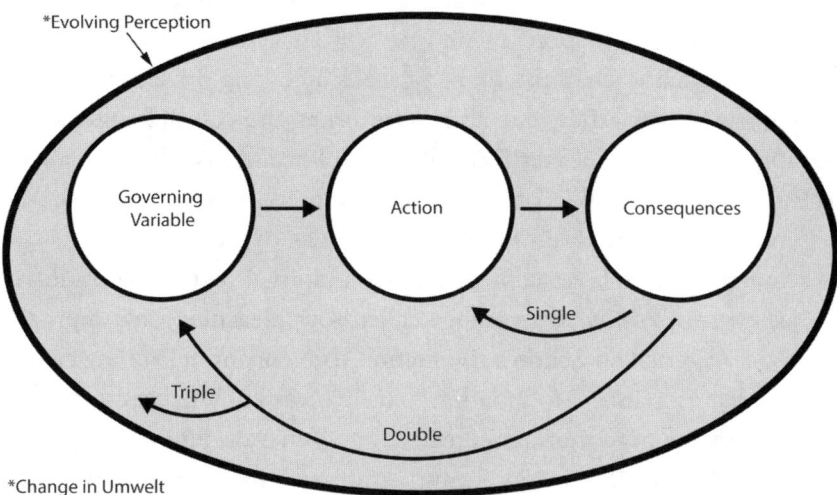

Figure 4.9. Single-, double-, and triple-loop approaches to learning result in fundamentally different outcomes. Single-loop learning essentially continues existing pathologies. Double-loop learning questions assumptions, but does so in the context of current paradigms. Triple-loop approaches are transformative by: asking if the fundamental approach is effective; potentially changing the entire approach to problem solving; changing the Umwelt/worldview of the participants.

approaches, as in the case of groundfishery management and the large landscape adaptive management programs reviewed in the next chapter.

By contrast, in double-loop learning, the entities involved assess the assumptions, policies, or values that led to their actions. This approach is similar to what Armitage[84] called experimental learning. In essence, it is learning by doing through an iterative cycle that involves four stages, including "concrete experience, reflective observation, abstract conceptualization, and active experimentation." The approach requires rethinking the underlying process and science and can have a strong role in providing the experimental framework to test alternative policy options. Armitage and colleagues called this a reflective process, in which the individuals' perceptions are altered. Much of the challenge of adaptive approaches to design rests in moving beyond single-loop approaches to explore alternative frameworks for problem solving. Adaptive management and ecological policy design considered in the next chapter can be understood as ways of incorporating double-loop learning into conservation and resource management. Double-loop learning is typically considered more resilient and sustainable than the more typical single-loop-driven "command and control" approaches, but also more time- and resource-intensive.[85] The approach in the near term can also be politically risky, for in being reflective, the shortcomings of current procedures are likely to be exposed and organizations are rarely willing to recognize or admit error.

More recently the concept of transformational or triple-loop learning has evolved.[86] In what Armitage called social learning, this approach builds on Argyris and Schön's foundation to incorporate the lower two levels of learning into a foundation that involves knowledge creation. It is coupled with organizational creativity to generate innovation within the organization. Flood and Romm examined triple-loop approaches through the lens of three interconnected questions: are we doing things right, are we doing the right things, and is the "right" approach leading to more effective action?[87] Triple-loop learning is transformational by affecting not just individual behavior and thinking, but also the identity or *umwelt*,[88] the worldview through which the organization sees itself within the overall context of the challenges it faces.[89]

An example of a transformational approach is the exchanges between Malpai ranchers and the Maasai (fig 4.1), in which both were facing periods of dramatic change and each learned through the lens of the other's experience.[90] In this *over-the-horizon learning* the ranchers were facing fragmentation and the loss of their land and livelihoods, whereas the Maasai, who now have comparatively large, open landscapes, are facing pressures to develop a system of land tenure that creates individual property rights. The exchange allowed each group to see the unanticipated outcomes of their actions. For the ranchers, seeing unfragmented landscapes was transformative in understanding how open spaces function, whereas for the Maasai, seeing the outcome of landscape subdivision was key to having them make more strategic land-use decisions. Out of the interchange, organizations such as SORALO (discussed in chapter 1) were born. These types of interchanges have been effective in a range of settings, from the long-term work of such organizations as the Quebec-Labrador Foundation, which for forty years has brought diverse groups together in Canada,

Central America, and especially the Middle East; and the Arava Institute for Environmental Studies and Seeds of Peace, which bring Middle Eastern youth together to craft nonviolent solutions to the region's social tensions. These interchanges also occur through more informal cultural exchange programs.

The essential point is that adaptive approaches to conservation or management that involve double, or better, triple-loop learning are the foundation of building resilient, durable processes for conservation science and policy design. This concept is fundamental because, as the case studies in the previous chapters demonstrated, sustainable, resilient approaches are impossible without these forms of reflective practice. Lessons from organizational theory demonstrate that people need to be open to exploring personal and organizational questions where learning takes place in a climate of openness in which political behavior is minimized or incorporated into the learning process.[91] But this approach is never easy. In the next section, we will transition to examining strategies that apply and integrate cognition with design to promote more effective conservation and reduce policy limitations.

Application of Theory and Practice

A primary goal for this volume is to develop more resilient and durable approaches to conserving large, complex systems through a reframing of science and policy. As discussed in the first chapter, one of the main places where social and ecological principles and processes intersect is within the context of commons. In these open, integrated spaces, social and ecological processes play out, and the interplay of these elements can be structured through strategic, proactive approaches. The concept of the commons is not intrinsically large-scale, but applies to any natural resource question at any scale that highlights the role of access, ownership, and governance in generating emergent approaches to conservation and management.

The paradox of the commons is that, in contrast to Hardin's premise in "The Tragedy of the Commons," where taken on face value the commons should decline, in practice they often are more durable than centralized governance systems (see discussion of commons in chap. 1). In what are now referred to as "panaceas," unified control of natural resource management, such as promoted by Hardin, often led to decline, rather than sustainability in resources. The decline occurs because most centralized governance is simply too slow, rigid, or unresponsive to the dynamics of social and ecological processes and therefore is intrinsically maladapted to respond to change and maintain complex systems over the long haul.[92]

So how is the paradox of the commons achieved? Effective commons institutions promote sustainability essentially not by generating stability, but through developing resilience.[93] This is done through the emergence of self-governing institutions that effectively link the environment with individualized and group cognition. In essence, effective governance generates, promotes, and sustains double- and triple-loop learning. Core to the theory are eight design principles that are prerequisites for durable common-pool resource arrangements, as presented in chapter 1—factors such as open process and aligning of scale (see fig. 1.7).

The long-term goal of efforts to promote effective science and policy design is recognition of which combinations of variables tend to lead to durable approaches and which lead to atrophy and collapse. One can image the decision-making space, as with the environment, as being a rugged landscape with peaks and valleys.[94] It is a thermodynamic world where the far-from-equilibrium solutions lie in domains that maximize the integrity of ecological and social systems. These solutions sustain ecological/evolutionary processes, as well as the richness of cultural and social elements, yet have requisite variety by providing more options in the physical and social domains and thus higher, far-from-equilibrium peaks. Thus, advances in social design lead to advances or new plateaus in function and opportunity in ecological systems too, while healthy ecological systems can maximize social opportunities. But as the Malpai borderlands example showed, these peaks are achieved only through considerable inputs of time and energy, and need broad bases of support, for if they are too narrow and steep, they are prone to collapse (fig. 4.10).

Conversely, social dysfunction is tightly coupled with ecological decline, leading to a downward trend in the ecological and social potential of the system. Agrawal in 2002 integrated the lessons from the commons literature to examine the critical enabling conditions for sustaining commons. These guiding principles, in essence, illustrate how good

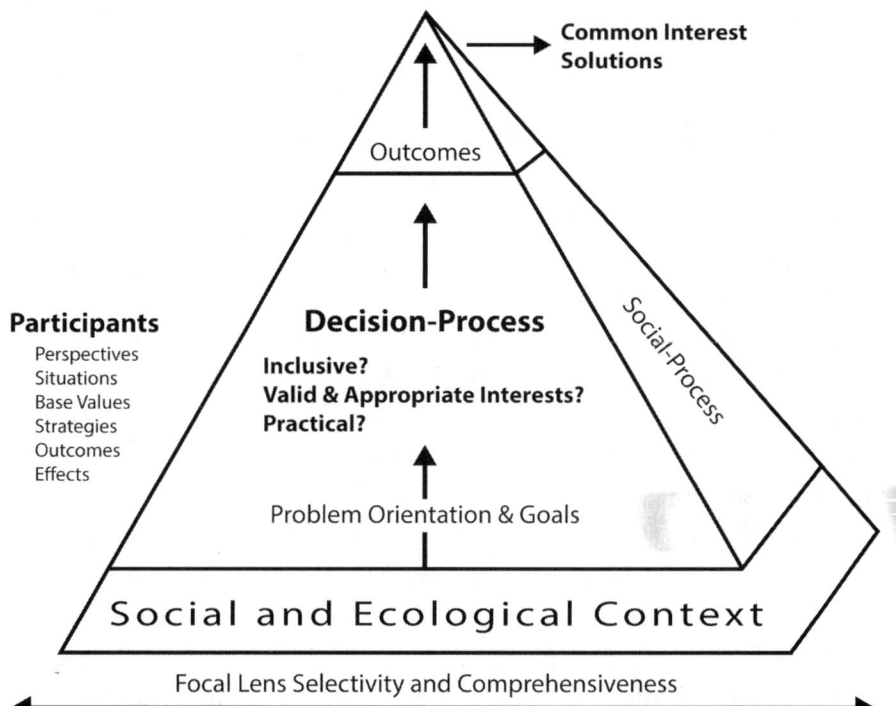

Figure 4.10. *The "sweat equity pyramid" depicts that for effective problem solving a broad base of understanding and trust needs to be developed and sustained. (Courtesy of Seth Wilson, the Blackfoot Challenge, Montana.)*

governance emerges from establishing the preconditions for success by examining the institutional properties that lead to good decision making.[95] The next section examines these principles in action through application of the concept of social innovation and the metric of adaptive capacity that pulls together the physical, ecological, social, and cognitive elements found in an integrated approach to developing more effective science and policy.

Social Innovation and Adaptive Capacity

Social innovation is an approach, process, or program that profoundly alters the basic routines, resources, and authority or beliefs in social systems.[96] These ideas are not new; they have been widely written on since the 1960s and extend back to Schumpeter's work in the 1940s, and beyond. Benjamin Franklin in the 1700s, for example, talked about social innovation in terms of small changes in the social organization of communities.[97] The capacity of a system to generate social innovations is an important contributor to overall system resilience. In the "paradox of agency," humans are deeply conditioned to the relative stability generated by social systems. At the same time, when social systems do not evolve, they become brittle and prone to failure as they fall more and more out of step with social and ecological realities.

Francis Westley, at the University of Waterloo in Ontario, Canada, spoke of social innovations having durability and broad impacts in a dynamic process that requires "both the emergence of opportunity and deliberate agency, and a connection between the two."[98] Key to this is seeing potential. The Canadian business theorist Gareth Morgan wrote an apocryphal story[99] of two children who were promised a great surprise waiting for them in a room behind a closed door. When the door was opened they found the room full of horse manure. One child started to cry. The other dove in, digging as fast as he could, exclaiming: "With all this manure there must be a pony in here!" Although there are cycles of opportunity, in the end, opportunities are created as much as they are encountered, and effective science and policy is all about creating the preconditions for generating opportunities for innovation, as well as recognizing them when they appear.

Adaptive capacity then is the ability of systems to respond to change. It relies to a large extent on the ability of a system to generate innovation. Achieving this durability is the emergent outcome of opportunity and deliberate agency based on design of governance and associated social processes. As a practical matter, change cannot occur unless a system has reached a level of stress where dramatic change is possible,[100] but also requires social innovators to serve as catalysts.[101] The formation of the Malpai Borderlands Group discussed in chapter 2 is an example in which all the processes and opportunities aligned at the genesis of the group. Although every situation and opportunity is different, a fundamental goal of a science of open spaces is to make such events less serendipitous and more a function of strategic processes design, points we will return to in the final chapter.

Concluding Remarks

We have reviewed the conceptual foundations of a science of open spaces that blends physical, ecological, and social theory. A fundamental tenet of this perspective for sustaining large and complex systems is to take a transdisciplinary approach that seeks a synthesis and synergy between a diversity of perspectives that are key to sustaining large systems. The phrase "open spaces" itself is intended to highlight the need to transcend the boundaries of academic disciplines and develop more integrated approaches that, like the systems they seek to sustain, are complex and emergent.

However, to design for sustainability and innovation, it is important to be aware of the cycles of change and the windows of opportunity that open and close within systems, those approaches that lead to durable science and policy, and those that intrinsically lead to failure. One branch of conservation and natural resource management is essentially the study of generating innovation and opportunity from the intrinsic cycles of collapse and renewal; this is the study of resilience. In the next chapter we review more than forty years of experience with resilience science, for it provides the crucial linkage between theory and practice that is necessary to sustain open spaces.

CHAPTER FIVE

Resilience and the Socioecological Synthesis

It is a riddle, wrapped in a mystery, inside an enigma; but perhaps there is a key.

—*Winston Churchill*

The introductory chapters and associated case studies illustrated the opportunities and inherent pathologies of preserving large, open spaces in rangeland and marine systems. The previous chapter grounded the lessons from these experiences in the underlying foundational theory. This chapter, through an examination of the paradigm of resilience, couples theory and practice through a review of more than forty years of effort to integrate social and ecological perspectives into a synergistic whole. This understanding of resilience is foundational to a science of open spaces, for it provides a context for the exploration of future opportunities for synthesis, and the foundations for the institutional design approach developed in the final chapter.

Foundations of the Resilience Paradigm

Beginning in the 1960s a wave of change swept across much of academia, as social revolution coincided with equally profound upheaval in the arts and sciences.[1] Fields as diverse as ecology, psychology, and physics began to explore more dynamic and emergent approaches to problem solving. Though the ways in which these approaches were undertaken varied within and across disciplines, one unifying theme common to physical, natural, and social sciences was the concept of resilience.

In psychology, resilience was defined as the ability of individuals to recover from trauma.[2] For example, psychological studies of children growing up in high-risk environments examined how some individuals survive and even capitalize on stress, whereas others founder. In risk management, resilience involves preserving the activities of human communities and societies, such as the mitigation of hurricane impacts or the ability to recover following earthquakes. Ecological resilience focuses on the long-term survival and functioning of individuals, populations, species, and ecosystems within variable environments.

Australian researchers Handmer and Dovers in 1996 distinguished between two fundamental perceptions of resilience: reactive and proactive. Reactive approaches seek only to return systems to the status quo, whereas proactive approaches are concerned with their sustainability through time. This typology extends across three distinct approaches to resilience: type 1—resistance and maintenance; type 2—change at the margins; and type 3—openness and adaptability.

Type 1 resilience, which seeks to achieve a high level of stability and constancy, sits at the reactive end of the spectrum. The groundfishery example is emblematic of this approach, in which extensive regulation seeks to control the negative impacts of overharvest. However, the fundamental mismatch in scale between the spawning aggregations of the fish and the size of the fishery leads to intrinsic degradation of the system.[3] Although new laws or changes in policy can mitigate some of the shortcomings of this approach, the focus on addressing the symptoms and not the cause of problems makes the fishery inherently unsustainable. Furthermore, in the type 1 approach, uncertainty about economic ramifications is often used as a justification to do nothing or trumps biological data, even when it is clear to both fishermen and regulators that the status quo only increases the dimensions and long-term costs of the problem. The groundfishery example is telling for the focus on discrete stocks and a council process of primarily financially or politically vested players. Coupling with centralized, expert-driven science provides a rigid political fix to a dynamic environmental problem. In this approach to fishery management financial hardship and uncertainty are frequently used to support decisions that fly in the face of the precautionary principle, which states that if an action or policy has a suspected risk of causing harm to the public or to the environment, in the absence of scientific consensus, it is wise to take the most conservative course of action.

Type 2 resilience embodies facets of both reactive and proactive approaches. The actors in the system recognize that the current approach is not sustainable and that new approaches are needed, but at the same time are unwilling or unable to undertake fundamental change. The danger of such half measures is that under the guise of doing good (albeit at a limited level), actors may simply be delaying the inevitable until the range of options becomes narrower and the cost of change more expensive. This phenomenon demonstrates that technological advances do not fundamentally change behaviors, they prolong them. Ecologist Tim Allen of the University of Wisconsin dubbed this the "Prius Effect." For example, cars with greater fuel efficiency may just delay the development of more effective mass transportation; if cars remained inefficient, unreliable, and expensive to operate, other alternatives would be implemented earlier. As it is, cars have become more efficient, but people drive more, so the net effect on the environment may not change as much as anticipated.

Unfortunately, this "change at the margins" approach is the most prevalent of the three in contemporary industrialized societies, where substantive changes are rarely effected deliberately. Again this is illustrated by existing groundfishery policy, whose primary tools are catch limits and management of fish stocks, because they are politically expedient and biologically simple to measure. Regulators, scientists, and even conservationists rarely, if ever, ask if these are even the right questions in the first place, or relevant factors to measure. The pathologies run deeper still, in that the fundamental premise of maximum sustained yield is profoundly flawed, because population numbers of fish are difficult to accurately assess and nearly impossible to measure until after the harvest when the actions are already done. As such, they are at best trailing indicators in service of management that needs to be proactive, and not reactive.

Type 3 resilience is the only truly proactive approach, as it addresses the underlying causes of environmental challenges. It seeks to redefine the dimensions of a given problem to find fundamentally new solutions. However, this approach frequently requires overriding vested interests to work for the common good. It also typically involves generating short-term costs to attain long-term benefits, which goes against much of human nature, political realities, and most institutional structures. And yet, with so many environmental and social problems, it is clear that the present frameworks are not working and that bold and innovative approaches are needed. A type 3 approach to mitigating climate change, for example, might include instituting measures to significantly reduce population growth, or redesigning cities to reduce the waste of suburbs that are predicated on cheap fuel and the convenience of cars.

Although tough to attain and sustain, there are certainly examples of type 3 approaches. The formation of the Malpai Borderlands Group and the organization of the Maine lobster fishery are both examples of this kind of thinking, in which change in social structure and governance has profound emergent outcomes. The Malpai group formed new and innovative alliances that allowed for proactive problem solving. Following the collapse of the lobster fishery in the 1920s, Maine lobster fishermen adopted innovative cultural norms that protected the breeding females and laws that required use of the double gauge, allowing harvest of only the medium-size animals, thereby protecting the juveniles, and the reproductively important older adults. In both cases these entities linked ecological constraints with social realities to form more resilient linked socioecological systems.[4]

Fundamental change also frequently means giving up some measure of control; some chaos is perhaps necessary to invite the institutional flexibility of an adaptive system. The American and French Revolutions, the Enlightenment, and the recent Arab Spring are all examples of axiomatic change that also illustrate the danger, volatility, and complexity that accompany such approaches, as well as why conventional forces resist countercultural movements. This change can take decades to achieve, as new generations who know only the new paradigm need to become established before the approach is ultimately successful, as happened with the lobster fishery, where what were once fairly radical ideas about resource stewardship are now strongly defended cultural norms. The three types of resilience have a close relationship with the three types of learning expressed in the previous chapter, illustrating how learning and institutional design are entwined. Type 1 resilience is akin to single-loop learning, type

2 embodies double-loop learning, and type 3 is essentially a triple-loop approach.

I present this framework to make the point that a science of open spaces is sought as a precondition for type 3 change. It is transformative in seeking an approach to research that is fundamentally different in asking not how to maximize production in peer-reviewed journals or forward the careers of researchers, but how to craft approaches to knowledge gathering that are more consistent with the scale and complexity of environmental problems while also addressing fundamental theory.[5] To successfully undertake current changes in science and policy, it is important to not reinvent the wheel, but to learn from the bold and innovative science and policy experiments begun in the early 1970s that sought to address comparable gaps between the potential and practice of conservation and management.

Ecological Resilience

Of the myriad perspectives on resilience, ecology's approach is most directly applicable to sustaining open spaces. The concept of ecological resilience, as developed in the 1970s, was a fundamental departure from earlier approaches to resource management in recognizing that rigidity often leads to brittle responses that are prone to collapse, so flexibility becomes key to successfully adapting to change and transcending environmental stresses. Ecological resilience-based perspectives do not look only at the ability of systems and their individual components to rebound following a perturbation, but more at their inherent ability to respond to and survive unforeseen events.[6]

In many ways, the framers of these definitions of resilience, though they did not use the same terms, were essentially interested in processes that sustain social or ecological systems through time in a perspective that hews closer to type 2 and 3 than type 1 approaches. The foundational papers in ecological resilience were similar to one another in their efforts to redefine the meaning of resilience from being reactive and rigid to proactive and emergent. Over the last four decades, the application of the ecology-based resilience paradigm has expanded to encompass such diverse fields as business, political science, urban planning, and many others. All are seeking answers to essentially the same question: how do we generate durable processes and approaches that embrace and capitalize on change?

The Canadian Connection

In the early 1970s, a group of scientists representing a range of disciplines in the natural sciences gathered at the University of British Columbia with the goal of remaking the fields of ecology and natural resource management. At the head of the Institute of Animal Resource Ecology was Crawford S. ("Buzz") Holling, a colorful and charismatic Canadian from Northern Ontario. Holling's informal conversation, often spiced with expletives, belied his upbringing on the frontier in the Far North and his sheer passion for innovation and research that addressed both applied and theoretical questions. He had already written some of the seminal works in mainstream population biology in the late 1950s and early 1960s before diverging into systems science, a path that look him increasingly far from the fold.[7] As it became apparent that conventional approaches to ecology were neither relevant nor sufficient for effective resource management, Holling and colleagues pioneered a new synthesis, one which would break down intellectual silos and link social and ecological perspectives, making theory more accessible to practitioners while using large-scale studies to test underlying theory.[8]

In many respects, the major contributions of Holling's Institute of Animal Resource Ecology were not just empirical or intellectual, but social. They promoted an innovative approach to science, explicitly working across disciplinary boundaries while remaining grounded in an ecological context. Their approach favored accuracy over precision, showing that a good question is more important than a trivial yet detailed answer. This interplay between theory and practice laid the foundation for what would evolve into resilience and sustainability studies. Its expression took three complementary but divergent forms: resilience, adaptive management, and ecological policy design (fig. 5.1).

Rise of Resilience

Ecologists and natural resource managers have long been preoccupied with understanding how ecosystems respond to and persist under natural and human disturbance, extending back to at least the work of George

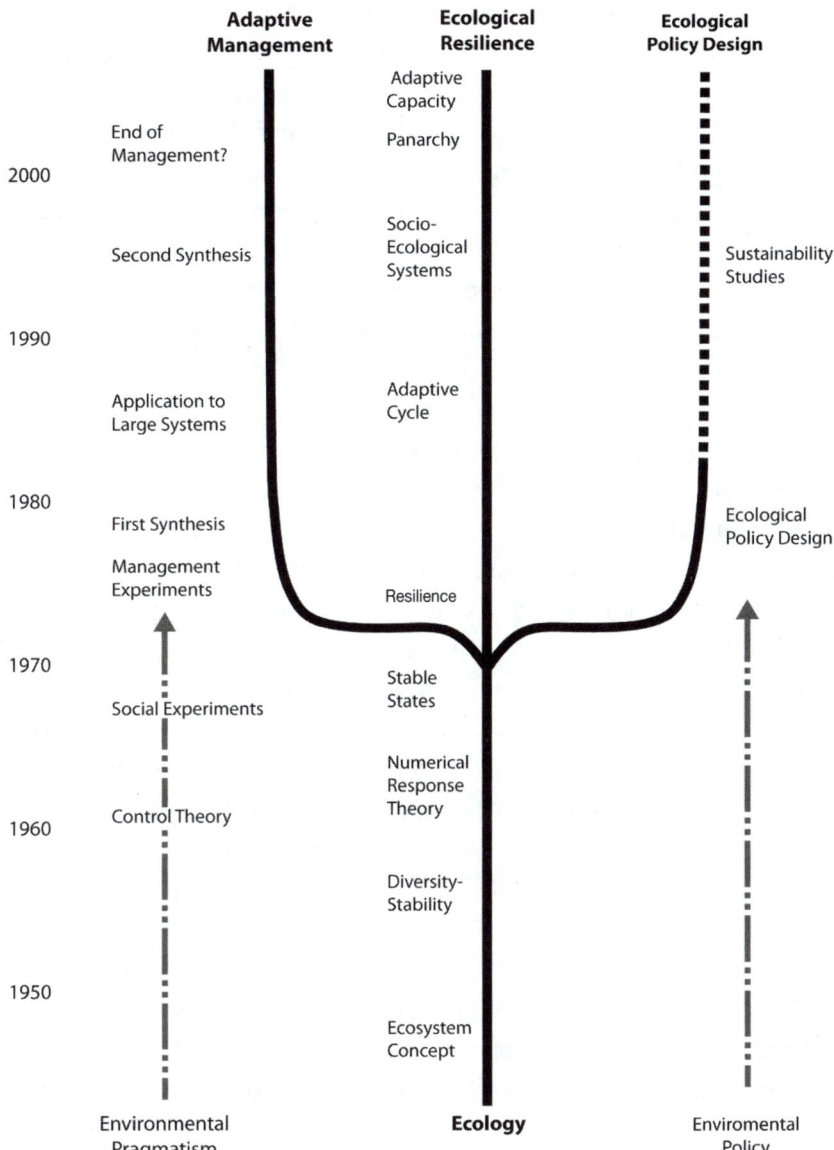

Figure 5.1. *A family tree of resilience-related concepts, tracking the key shifts in thought through time. Resilience science emerges from three core bodies of thought: ecology, environmental pragmatism, and environmental policy. Each has its own discrete but complementary perspectives. (After Curtin and Parker, 2014.)*

Perkins Marsh in the 1860s. Conservation pioneer Aldo Leopold contributed to many of the foundations of resilience, as evidenced in the small collection of essays known as *A Sand County Almanac.*[9] Especially relevant is the book's closing chapter titled "A Land Ethic," in which Leopold laid out his vision for interdisciplinary synthesis in conservation by presenting a complex, systems-based view of what today we call socioecological systems. The chapter recognized the role of values and that for effective conservation, landowners and practitioners cannot just derive the maximum production from a natural resource, but must also find a "gentler and more objective criteria for its successful use." The criteria, or "land ethic," essentially valued long-term form and function in an approach that was neither utilitarian nor preservationist, but sought a dynamic balance and synthesis between these two often competing worldviews.

The question of sustaining ecological function under natural and human disturbance became a major focus of academic ecology in the 1950s. In what became known as the "diversity-stability" hypothesis, Robert MacArthur postulated in a 1955 paper that community stability was largely tied to the pattern of interconnectedness of food webs. The focus on stability as a cornerstone of ecology achieved a high-water mark at the Brookhaven Symposium on "Diversity and Stability in Ecological Systems" in 1969. The most influential paper of the symposium was a short work by Harvard evolutionary ecologist and geneticist Richard Lewontin titled "The Meaning of Stability." Lewontin's paper proposed a "vector field model" to describe the interaction of stability and community structure in which he made an almost passing reference to "basins of attraction" that generate stability within a system of "alternative stable states." Though the vector field model is largely forgotten, the "basins of attraction" concept, in which systems can move between alternative stable states ("basins"), rather than along a single, simple trajectory, formed a foundation of what was to become resilience science. Under this paradigm, resilience is the capacity of a system to flip between alternative states or to rebound following disturbance.

By the early 1970s, two major papers challenged the primacy of stability as an organizing principle in ecology, conservation, and management. The first was Australian physicist Robert May's theoretical work in 1972, mathematically demonstrating that stability was not intrinsically linked to diversity. The other was Holling's monograph "Resilience and Stability of Ecological Systems," in 1973, which eliminated the ambiguity in the "diversity versus stability" debate. He proposed to get ecology out of the theoretical "stability" rut by illustrating that it was only a part of the larger question of how to sustain ecological systems. As a concept, stability was used to examine both a system's behavior near equilibrium and its long-term potential for persistence, two distinct questions. Rather than addressing these concepts as one under a focus on stability, Holling suggested dividing it into two principles: resilience and stability. Holling pointed out that a system can fluctuate greatly (i.e., have low initial stability) and still be extremely resilient (fig. 5.2).

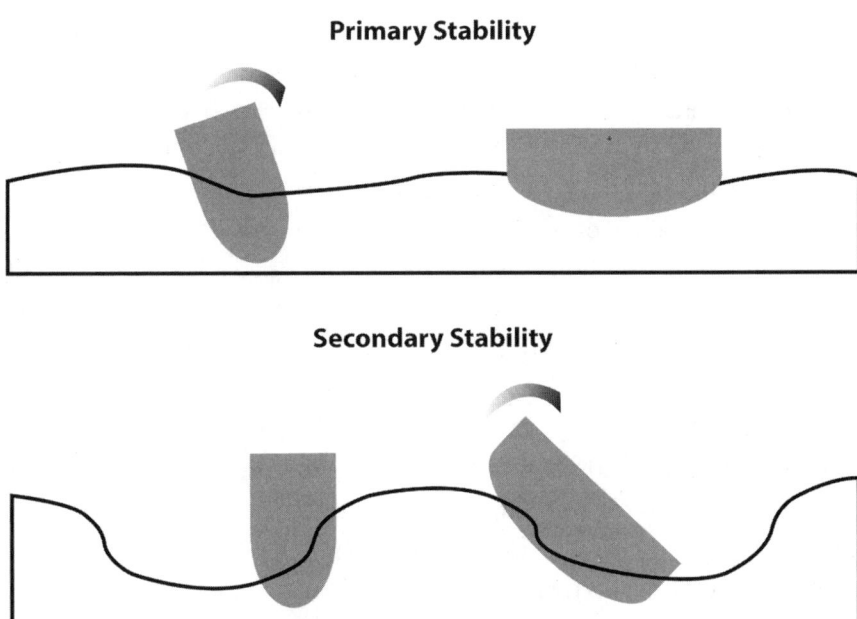

Primary Stability

Secondary Stability

Figure 5.2. *This depiction of primary and secondary stability of two canoe hulls illustrates the relationship between stability and resilience. The hull at right has good primary stability (i.e., it is a very "stable" platform, but relatively unseaworthy under rough or changing conditions, not very resilient). The hull at left has good secondary stability. It will seem more unstable or "tippy" initially, but the more curved hull is actually more seaworthy (i.e., resilient) in rough seas, because it is better able to move with the water and respond to turbulence without tipping over.*

Holling stated: "The stability view emphasizes the equilibrium, the maintenance of a predictable world, and the harvesting of nature's excess production with as little fluctuation as possible" (i.e., the maximum sustained yield approach favored in fisheries and forestry, wherein managers attempt a consistent high yield of resources). The resilience view, in contrast, emphasizes the domains of basins of attraction, alternative stable states, and the need for persistence under varying levels of unpredictable disturbance. Holling next laid out the tenant that became transformative in the management of natural resources: a focus on increasing stability frequently reduces the resilience of the system. The very process of trying to maintain a system within a narrow range of limits may actually increase the likelihood of its collapse. In contrast to stability-driven approaches, which seek to control systems within human-defined bounds, resilience is persistence-driven, focusing on longevity-inducing behaviors that maintain a system while withstanding perturbations. Resilience management begins with an assumption of insufficient knowledge of the complex dynamics of natural systems and employs the guiding principle "expect the unexpected" regarding these systems' responses to change.

The next major conceptual breakthrough in the evolution of resilience came in a 1986 book chapter by Holling titled "The Resilience of Terrestrial Ecosystems: Local Surprise and Global Change." The key concept to emerge from this work was the "adaptive cycle" (fig. 5.3).

In developing the adaptive cycle model, Holling built upon the classic concept of ecological succession in natural communities that was forwarded by Clements in 1916. Yet, in resilience theory, succession was only a part of the larger cycle that included four distinct yet interconnected phases. During the initial "exploitation" phase, the system experiences rapid new growth (succession), as seen in a forest following clearing. This period of growth leads to the "conservation" phase as resources accumulate and eventually become more concentrated in ever tighter connections, building toward an inevitable threshold of dramatic change. The system becomes so overconnected that the resources are suddenly released in a period of abrupt transformation, during which the system collapses. "Surprise" events during the "release" phase are an intrinsic part of long-term cycles, as systems cross thresholds during transitions in

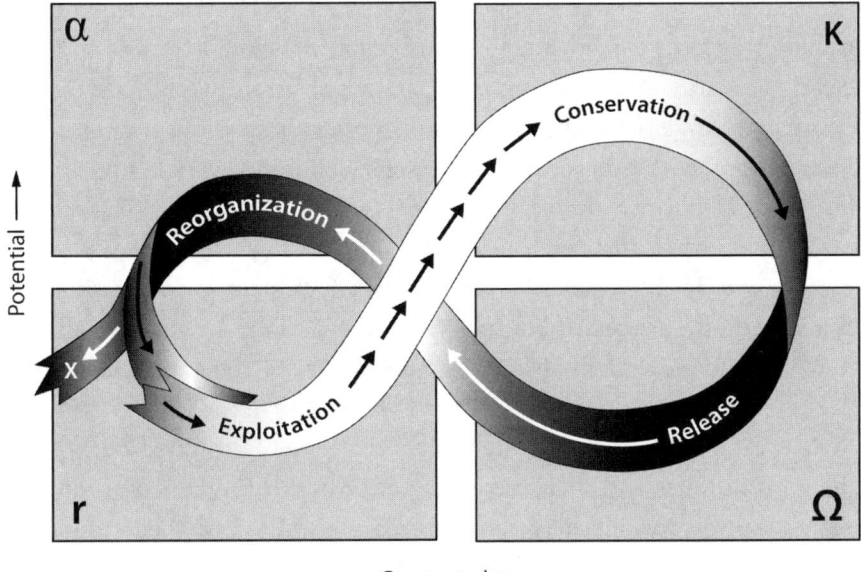

Figure 5.3. *The adaptive cycle depicting the changing dynamics of systems that are intrinsic to ecological and social systems. The beauty of the model is its broad applicability, providing practitioners a context within which to consider their actions. For example, in the conservation phase, opportunities for change may be limited, whereas in the reorganization phase, there are more potential opportunities for new and innovative solutions. (After Holling, 1986, and Gunderson and Holling, 2002.)*

structure. In natural systems, pest outbreaks, disease, or catastrophic fire are all the inevitable outcomes of a system and its resources becoming overconnected. Finally, in the "reorganization" phase, the system either continues the adaptive cycle by making the resources available again or it takes a different path altogether (such as a forest becoming a grassland following a fire).

Viewed from a social perspective, the model has significant implications for understanding cycles of opportunity in organizations and society as a whole. For example, in fisheries or ranching, when people are deriving substantial incomes, change is all but impossible, for there is little incentive to undertake actions that are economically and socially expensive or risky, even if current actions are widely know to be intrinsically unsustainable (as in the groundfishery example). However, when communities and the system they live in pass a threshold of stress and it is clear the status quo is failing, they are much more likely to seek and embrace change. This cycle of creative opportunity is not new, but was first recognized by the economist Joseph Schumpeter in 1942. In what he termed "creative destruction," events such as market collapses are negative in the short term, but positive in the longer term because they provide windows of opportunity during which people and institutions fundamentally reassess their options and undertake new opportunities, much as the New Deal and other transformative social policies of the 1930s were an outcome of the Great Depression. In the case of both the Malpai ranchers and the Maine coastal fishermen, profound changes in the communities and their ecosystem allowed for fundamental rethinking of the relationship between these communities and their resources. Holling and Schumpeter's point is that these events are not isolated periods of misfortune and opportunity, but intrinsic components of all systems. So in looking for opportunities to make a difference, practitioners need to seek out communities and resource bases that are ripe for change, and not try to force change when a system is not ready.

The model is powerful in both its relative simplicity and its ability to integrate existing concepts of stability and resilience by making them relevant to ecology while bringing in social perspectives from economics and psychology. The adaptive cycle combines both equilibrium and non-equilibrium approaches to show that patterns and processes appear stable or unstable often as a reflection of the scale at which they are viewed. The smaller the scale, the more unstable systems tend to be. This has fundamental implications for conservation, for it demonstrates from a perspective other than that of the species-area curve why large-scale approaches are necessary.

In understanding resilience, the degree of connectedness becomes the key: underconnected systems may be unable to respond to a change or threat, and overconnected systems may seem more stable but are actually more brittle and prone to sudden collapse. The paradigm that stability can actually undermine resilience challenged a fundamental ecological assumption by suggesting that more complex systems may appear more stable in the short term, but depending on the extent of connectivity, may be less resilient in the long term. Take the example of a coral reef or tropical rain forest. Although the higher species diversity may make it less prone than a less diverse ecosystem to significant changes at the community level, so as to appear more stable, the tight interconnectedness of the whole system may make it much more vulnerable to wholesale collapse from large perturbations such as climate change.[10]

The adaptive cycle also demonstrated the power of metaphor in scientific modeling. That good ideas need not contain calculus or differential equations has been the bread and butter of theoretical ecologists since Robert MacArthur's time. The use of descriptive models and analogy can often be more powerful in communicating abstract ideas. This descriptive approach to the heuristics of resilience has continued through to the present, perhaps explaining the concept's wide use across a range of disciplines.

Holling's 1986 chapter made a clear leap from resilience as an essentially ecological concept to one that had relevance beyond resource management to a range of human systems. In the 1990s, many of the major contributions revolved around the integration of social and ecological variables, evolving primarily through the work of Fikret Berkes, Carl Folke, and colleagues. In 1993, at the Beijer Institute of Ecological Economics in Stockholm, these researchers developed a common framework to link social and ecological approaches, which led to their 1998 edited volume *Linking Social and Ecological Systems*. In the book, they used case studies to refine social ways of problem solving within an ecological context. The essential goal was to create a transdisciplinary framework to evaluate examples of socially and culturally evolved management practices. The authors addressed two common objectives: (1) how the local social system has developed management practices for dealing with the dynamics of the ecosystem(s) in which it is embedded; and (2) how the social mechanism behind these management practices promotes or decreases resilience and sustainability.

To meet these objectives, the book explores three hypotheses. First, maintaining resilience is important for both resources and institutions. The entwined nature of the sustainability of social and ecological systems suggests that both must be considered as an integrated whole, just as many of the successful examples of natural science in this book are a product of the social context within which they are embedded.

Second, resilient resource management promotes disturbance at specific scales to prevent the disruption of the structure and function of the overall system and to contribute to its long-term persistence. This is analogous to prescribed fire's reduction of fuel loads through a small loss of biomass in the near term and resultant decrease in the likelihood of a catastrophic fire in the future.

Third, the social mechanisms behind these management practices exhibit a coevolutionary relationship between local institutions and the ecosystems within which they are embedded, much as in the examples of fire management in the borderlands, Maasai grazing, or community-level management of lobster, where the feedbacks between social and ecological processes and constraints profoundly influence the structure of each. Together, these practices and mechanisms provide a reservoir of active adaptations to establish an interactive process by which resilience and sustainability are maintained through time.

Berkes and Folke's book employed resilience theory as a means for rethinking the underlying assumptions of resource management. Resilience theory applies a systems approach to find common denominators that link social and ecological perspectives and that lead to simple rules that promote emergent sustainability-promoting outcomes. This happened with the Kenyan pastoralists and Maine fishermen; the resources they rely on were sustained when effective linkages were found between ecological and social variables that integrated vast amounts of complexity into relatively discrete and assessable action steps, such as daily milkings or catch records. Disconnects between social systems and natural processes and the lack of clear integrating variables, as in the case of the groundfishery, lead to almost certain system collapse.

In the early 2000s, resilience reached its most recent expression in the book *Panarchy*, edited by Lance Gunderson and Crawford Holling.[11] *Panarchy*, named for the cross-scale, crossdisciplinary, and dynamic nature of complex systems, drew upon the Greek god Pan to capture an image of unpredictable change in hierarchies across scales.[12] The book evolved from a series of five international meetings that occurred on islands around the world, from Little St. Simons Island off the coast of Georgia to Heron Island in Australia's Great Barrier Reef. The authors met as the "Resilience Network" to explore themes that had emerged from the previous three decades of work. The concept of panarchy provided an organizing framework for discussing complex dynamics (fig. 5.4) and how concepts of ecological resilience feed into creating more sustainable social and ecological systems.[13]

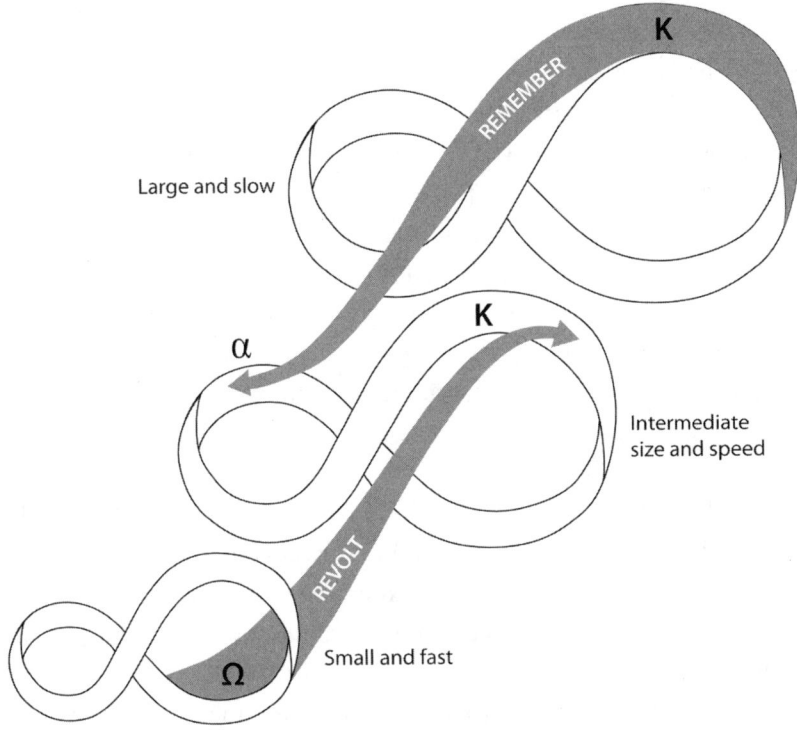

Figure 5.4. *Panarchies contain nested sets of adaptive cycles within larger hierarchical structures (from Gunderson and Holling, 2002). They show the link between connectedness and potential resilience across scales and how systems contain not just multiple scales, but also multiple rates of function, with small, fast facets of the system interacting with large, slow components. Multiple scales and rates of function create complex system dynamics and challenges with predicting or understanding the function of such systems, which can lead to surprise outcomes for conservation and resource management. Prediction is especially difficult at the middle scales where small and large processes intersect, and where most human action occurs.*

The core approach of panarchy is a departure from earlier resilience studies in drawing to a greater extent from complexity theory. In implementing the approach, Holling stated, "be as simple as possible, but no simpler than necessary." Minimal complexity requires (1) three to five interacting components, (2) three qualitatively different speeds, (3) nonlinear causation and multistable behavior, (4) resilience and vulnerability that change with slow variables, (5) biota that create processes that reinforce the structure of the system, and (6) self-organization resulting from spatial contagion and biotic legacies.

But what does this mean for practice on the ground? Though the concepts of panarchy have been widely cited for more than a decade, the sheer complexity and social design elements needed mean this framework is rarely achieved in practice. A science of open spaces in many respects seeks to operationalize the approach through considering complexity explicitly in the process of program design. The Malpai science program, though devised before the publication of *Panarchy*, essentially operationized the book's cross-scale, crossdisciplinary approach. The greatest insights came from viewing the interactions of multiple variables, across multiple scales, rather than considering them in isolation, with interactions among climate, fire, and grazing key to understanding the ecological function of the borderlands.[14]

The fundamental principles of panarchy theory proposed to (1) eliminate destructive constraints and inhibitions on system dynamics, (2) protect and preserve the accumulated experience upon which the necessary tools for change are embedded, (3) stimulate innovation through a range of safe-to-fail experiments that probe the most effective and durable ways forward with the lowest cost to individual careers and organizational budgets, and (4) encourage foundations for renewal that build and sustain the capacity for resilience in human and natural systems. These principles nicely summarize what we found out independently through the Malpai experience, as well those in fisheries and other arenas where both ranchers and fishermen strove to break down institutional barriers with the science providing not only data, but also common ground to bridge very different perspectives and cultures. Studies such as McKinney Flats in the borderlands were essentially safe-to-fail experiments that allowed us to learn from the outcomes of drought. By contrast, rangelands in Kenya exhibit the profound costs of policy failure, as illustrated by the dramatic wildlife die-offs. Although the principles of panarchy are ubiquitous, the devil is in the details, with a science of open spaces seeking to both put more meat on the bones of resilience theory and move us from theory to practice. The next section transitions from the foundations of resilience to exploration of the practical implications of theory with adaptive management asking the more pragmatic question: How can resilience be achieved in practice?

Adaptive Management

Although resilience theory and adaptive management are sister disciplines, there are some fundamental differences that can be traced back to their origins. The roots of adaptive management can also be attributed to Aldo Leopold, who advocated such an approach as early as the 1920s. In many respects, Leopold can be considered the "father of adaptive management." However, his ideas are embedded in a broader and deeper tradition of American thought that can be traced back to the pragmatists around 1900.[15] In the context of conservation, the pragmatists sought a middle ground between the romantic approach to nature embodied by John Muir and the utilitarian approach of Gifford Pinchot, by advocating an ethics-based view of the land and the sustainable use of its natural resources.

Another stream of thought that contributed to the adaptive management paradigm was the concept of action research,[16] which includes the research process itself as a part of a study. So in addition to gathering data, researchers also began self-assessing the impacts and efficacy of their methods, recognizing that the research process itself can be as informative as the results. The approach we took to experimental science in the borderlands was very much an intellectual descendant of action research, wherein we recognized that the social lessons learned from developing large-scale conservation science were every bit as important as the actual biological data derived from the studies.

In the 1960s, further intellectual foundations of contemporary adaptive management were laid down through developments in information theory. One of its antecedents was dual control theory, as proposed by Fel'dbaum in 1960 and 1961, in which systems whose characteristics are initially unknown were experimented with to both understand their behavior and control their outcome.[17] The goal of dual control theory was to develop a framework within which one could probe a system and look for responses without compromising it, much as experiments such as McKinney Flats were controlled microcosms of the ecological and social influences on much larger landscapes. Adaptive management was also influenced by an "experimental approach to social reform," popularized in an influential paper by social scientist Donald Campbell in 1969, in which social policies were considered experiments to be learned from, and not ends unto themselves.

The first paper to explicitly use the term "adaptive management" was a 1975 study of fisheries by Carl Walters and Ray Hilborn. However, Holling and Chambers's 1973 paper "Resource Science: The Nurture of the Infant" embodied the approach, borrowing heavily from the conventional ecological theory of the time to develop testable models through computer simulations of policy outcomes. What remains striking about these early writings is how rapidly they identified the process of applying adaptive approaches, a framework that has existed and been little improved upon to this day. In his 2006 memoir Holling noted:

> One advance developed a sequence of workshop techniques so that we could work with experts to develop alternative explanatory models and suggestive policies. We learned an immense amount from the first experiment. That focused on the beautiful Gulf Islands, an archipelago off the coast of Vancouver. We chose to develop a recreational land simulation of recreational property. I knew little about land speculation, but we made up a marvelous scheme that used the predation equations as the foundation—the land of various classes were the "prey," speculators were the "predators" and a highest bidder auction cleared the market each year. The equations were modifications of the general predation equations.

In addition to the simulation model (known by the acronym GIRLS, for Gulf Island Recreation Land Simulation), Holling and Chambers were ahead of their time in focusing on the role of interpersonal dynamics in effective decision making. To look at their paper today, one would never

Benevolent Despot · Peerless Leader · Snively Whiplash

Figure 5.5. *Three cartoons depicting characters that often appear in the policy arena. Their approaches capture a portion of the underlying dynamics of social systems with distinct and predictable pathologies recurring across different projects and programs (After Holling and Chambers, 1973).*

know it was written more than forty years ago. Their perceptions are illustrated in a series of cartoons (fig. 5.5) that identifies key personality types who engage in collaborative planning and adaptive management, from "benevolent despots," who "balance delicately between humane omniscience and programmed (or real) stupidity," to the "peerless leader," who sacrifices for the greater good, and the "snively whiplash," who detests and seeks to undermine the whole process. Perhaps the key figure described by Holling and Chambers is the "Blunt Scot," whose "bluntness and sincerity of purpose transcend the mischievous irresponsibility that most of the rest of us succumb to occasionally." The point was not to make fun, but to recognize that in collective decision making certain human traits emerge that inform the policy process.

These depictions of personality types are remarkably accurate, for I have seen most or all of them appear repeatedly in different policy arenas. From the borderlands to fisheries, and even in work with the Maasai and the Middle East, I see these same archetypes of characters appear with remarkable frequency. I have learned to look for them and recognize that they will all play their own role. I have also learned the importance of where possible transforming the critics and fence-sitters to proponents. For if someone can reach and engage them, then much of the rest of the process falls into place.

Holling and Chambers took the process of developing collaborative frameworks seriously, but also recognized that these decision-making arenas are essentially games (high-stakes games, but games nonetheless) with discrete rules and dynamic outcomes embedded in an intrinsically adaptive approach. This approach has framed the process of adaptive management in the decades since. Ray Hilborn,[18] one of adaptive management's early proponents, wrote:

> The term adaptive management has come to have several meanings, most wider than we originally intended as we were developing the concept in the mid-1970s. At that time, we were deeply involved in development of complex computer simulation models for a wide variety of resource and environmental management problems. As we developed more case studies, a common theme emerged: there were always critical gaps in understanding and quantitative uncertainty about key ecological processes that are sure to be important in determining outcomes of management alternatives (like the relationship between spawning and recruitment in fish), yet these processes had defied conventional scientific (experimental) analysis because they unfold on space and time scales that are very difficult to study. So our response to these uncertainties was to suggest that the management process itself should be treated as inherently experimental, involving judicious choices of policy actions that allow direct assessment (via the basic experimental concept of treatment comparisons) of what works best.

However, adaptive approaches are also not panaceas for experimental probing of systems, or direct experimentation, which is fraught with risk at a number of levels. As illustrated by the Malpai example, despite a decade of investment by researchers in experimental science in what seemed like a best-case scenario, the program still fell short of its potential. This resulted in considerable professional and personal cost to the scientists involved and demonstrated that experimentation is intrinsically risky. A balance between learning and risk lies at the heart of the optimal control approach as it applies to adaptive management, and yet, in situations of extreme uncertainty, there may be no optimal policy at all. At the same time, experience is both expensive to acquire and rarely organized

in a manner that maximizes the ability to anticipate and effectively respond to future events. Therefore, the challenge is to make the most informed decisions in the face of uncertainty. Adaptive management seeks
to develop an organized process to do this.

However, the drawback of adaptive management as currently practiced is that it uses past behavior as an experimental control against
which to compare the impact of future outcomes. Such a retrospective
approach is akin to driving down the road while looking only in the rearview mirror (fig. 5.6): you can't see what is coming until after you have hit
it. Monitoring with incremental revision is often not effective in a world
filled with thresholds. In the case of the East African droughts and wildlife die-offs depicted in the opening chapter, it was not until the system
was profoundly stressed that the real costs of more rigid conservation
became apparent. But by the time the costs were widely evident, it was
too late for an effective and timely response.

Accordingly, we must instead develop institutional structures that allow us to both learn and apply resulting lessons within a framework that
promotes continual analysis, revision, and action as in the learning frameworks explored in the previous chapter. This is the crux of a science of

Figure 5.6. *Monitoring is akin to driving only with a rearview mirror in that
you see what you have passed through, but not what you are about to hit. It is
important to incorporate retrospective (monitoring) and prospective (research)
approaches to conservation and management, yet few programs also incorporate
prospective approaches that typically include both modeling and experimental
testing of assumptions.*

open spaces that recognizes that not just passive measurement is needed, but also active probing of systems combined with critical analysis and application of the resulting lessons.

In the mid-1970s Holling and members from the University of British Columbia group moved their work to Laxenburg, Austria, where he became director of the International Institute for Applied Systems Analysis. Thus began a remarkable period of field trials in adaptive approaches that culminated in the seminal book *Adaptive Management*.[19] The institute's early transboundary experiments in sustaining open spaces consisted of nine projects featuring different levels of complexity, data, scale, and understanding, ranging from the aforementioned Gulf Islands project to studies of oil shale in the American West, spruce budworm in Canada, and capybara in Venezuela. Especially innovative was their use of divergent approaches in projects to gain a broader understanding of overall adaptive processes.

The most cited and closely followed of the institute's case studies was of the town of Obergurgl in the Tyrolean Alps (fig. 5.7). Obergurgl was an interdisciplinary project conducted as part of UNESCO's Man and the Biosphere program, "Study of Human Impacts in Mountain and Tundra Ecosystems." During an intensive, five-day workshop in 1974, participants from a variety of interest groups, including local residents and scientists, contributed input to computer simulations of planning options for the community.

At the time, Obergurgl faced a significant social problem stemming from tourists who came to the region both for its *Kulturlandschaft* ("cultural landscape"), with its quaint farms and open fields, as well as to ski. Obergurgl's cultural artifacts existed in direct conflict with the ski industry, which was planning to expand. Skiing on area slopes compacted the snow, resulting in slower melting rates and a delay in the growing season for agriculture. Ski industry leaders also pushed for the removal of fencing and other farming-related structures during the winter on communal lands near the town, which would protect skiers from dangerous obstructions, but disrupt agricultural production and add costs to farmers. Meanwhile, during the summer growing season, hikers caused erosion and interrupted farming activities.

These issues resulted in an uneven distribution of costs and benefits;

Figure 5.7. *The town of Obergurgl in the Tyrolean Alps was one of the first areas to which adaptive approaches were applied to link social-ecological systems. A UNESCO-based study illustrated how intensive workshops embodying an adaptive approach and modeling can have tangible benefits that are evident years later (e.g., Holling, 1978; Moser and Moser, 1986). (Photo courtesy of Shutterfly.)*

the few families running ski lodges and tourist centers received most of the benefits of the tourism while the farmers bore most of the costs. Although tourists were attracted to the region because of its pastoral landscape, the farmers who sustained the landscape did not receive the benefits generated by tourism. The Obergurgl project strived for a redistribution of the community's wealth by having the tourism industry partially subsidize farmers to compensate for the ecological costs of tourism, conserving both long-term land uses and open landscapes.[20]

Partly because of the UNESCO project, Obergurglers began to sense the importance of collective responsibility and the necessity of community action to ensure a sustainable future. Researchers followed the progress of Obergurgl for years thereafter. The program illustrated how adaptive management can move beyond focusing on a single resource to consider wider issues of social and ecological interactions, foreshadowing later approaches adopted by a wide variety of projects such as those in the Florida Everglades and Chesapeake Bay. Though Obergurgl was an isolated case, its overarching lesson is widely transferable: intervention that incorporates local knowledge and science-based approaches promotes development of the common ground and social capital that is essential for sustainable resource use. Such a legacy can clearly be seen in the Malpai Borderlands Group (MBG) approach nearly two decades later, where the insights from the ranchers' question of how the environment was changing, and what to do about it, led to active experimentation at a range of scales, from the intensive 9,000-acre McKinney Flats experiment examining climate, grazing, and fire interactions, to the 48,000-acre Baker II burn that was the largest prescribed burn ever undertaken in the United States.[21]

By the early 1990s, practitioners had enough experience with adaptive management in large-scale projects to begin to extract a wider array of lessons. One of the most influential writings of the time was Kai Lee's *Compass and Gyroscope,* published in 1993. Lee's book contributed social theory and practical experience from the policy realm, approaching adaptive management from a different perspective than earlier writings that were grounded primarily in an ecological context.[22] Lee wrote, "Linking science and human purpose, adaptive management serves as a compass for us to use in searching for a sustainable future" (p. 9). He also wrote, "Bounded conflict—the gyroscope—is a pragmatic application of politics that protects the adaptive process by disciplining the discord of unavoidable error" (p. 11).

Lee drew primarily on experience from the Columbia River basin, where adaptive management was applied beginning in 1984. He pointed out that social learning is most needed and applicable in large ecosystems where simple cause and effect solutions do not apply (the concept of "large" is not necessarily a definition of physical size, but encompasses the number of interacting social and ecological variables). He further described "arenas of interdependence" and "laboratories of institutional invention" where there is the potential for observing cumulative and emergent effects not visible at smaller scales. Lee conceptualized management as an experiment that had important sociological consequences where surprising results are seen as legitimate, rather than as evidence of failure, giving both managers and policy makers the freedom to take the proactive (and sometimes risky) approaches necessary for progress. The crux of Lee's book in many respects is that core adaptive management programs are, intentionally or not, experiments; therefore, it is imperative that we "learn from them."[23]

Lee distilled what he learned from the Columbia River and related projects into what he called "institutional conditions favoring adaptive management." These are, essentially, principles for effective research and policy design that are especially relevant to conserving large, open spaces:

- There is a mandate to take action in the face of uncertainty.
- Decision makers are aware that they are experimenting, whether they choose to or not, so they may as well learn from the experience.
- Decision makers need to care about improving outcomes over biologically relevant time scales (as opposed to political time scales).
- Testable hypotheses need to be formulated, and resources are sufficient to measure outcomes at scales relevant to conservation and management.
- The organizational culture must encourage learning from experience, and institutional design must be in place to respond to experience.
- There must be sufficient durability in the system, and institutional patience, to sustain long-term measurement at scales relevant to environmental and societal change.

These are all lessons shared in both the rangelands and fisheries examples. A crucial hallmark of the initial MBG approach was to develop a joint vision of landscape-level processes (i.e., climate, fire, and grazing) that also served as a testable model of ecosystem function, with experimental and monitoring data used to test and refine the assumptions of ecological function held by ranching and research communities (e.g., fig. 2.4). The decline in the science program was reflected in a decline in the institutional design that allowed for learning and adaptation.

Lee described four guidelines for harnessing social capital, including (1) wait for crisis, (2) take advantage of disorder and slack, (3) be skeptical of the value of information, and (4) be patient. The examples from the Malpai borderlands, the Maine coast, and East Africa illustrated these principles; a period of conflict and threat generated the preconditions that forced people to consider innovative alternatives to the status quo. As suggested by the adaptive cycle, these windows of opportunity are relatively short transitional moments in time, so practitioners need to recognize and capitalize on these periods of opportunity.

An implication of Gunderson and Holling's metaphor of panarchies is that events and opportunities occur not only across different scales, they are also distributed through time and space. Much as metapopulation models in ecology posit that populations of plants and animals wink in and out across the landscape, so too do periods of social opportunity. The implications of this are that to influence innovation one cannot stick always with one system in hopes that a period of transition (Holling's "release" phase in the adaptive cycle) will come along to allow for a transformative period. Instead, one must cast a broad net and look for windows of opportunity across a range of projects, while being objective in assessing if a program or location is in a stage where it can embrace transformative change.

So it was in the chronology of the fisheries and rangelands examples considered in this book. As the Malpai Borderlands Group project reached a plateau and period of relative stasis, the marine examples entered a period of innovation. The marine systems now appear to have reached a plateau, so in my own work I have transferred back to looking at western landscapes.[24] However, in contrast to Lee's Columbia River example, which was primarily driven by large government programs, the experiments presented here in a science of open spaces are emergent and locally driven, for as we will see in this chapter, the large, federally derived collaborative approaches are not the panacea anticipated in Lee's writing.

The next synthesis of lessons from adaptive management arrived in 1995, via *Barriers and Bridges*. The brainchild of systems thinker Stephen Light (with coauthors Gunderson and Holling), the book was designed to incorporate practical experience and theory. Its title refers to barriers to progress and the use of adaptive approaches as "bridges" to transcend these barriers. Like Holling and colleagues' 1978 adaptive management volume, the book relies heavily on case studies and transboundary comparisons from field trials in large, open systems. *Barriers and Bridges* was the result of a three-year research project, including three intensive workshops with the then-prospective authors. The book was designed to explore ways of dealing with uncertainty in the management of complex regional ecosystems through a focus on two key questions: (1) Do institutions learn, and if so, how? and (2) How do ecosystems respond to management actions? This focus integrates the lessons of the previous three adaptive management volumes[25] to link ecological and social theory with "empirical practice" and insights that were a prelude to Gunderson and Holling's 2002 book *Panarchy*.[26]

One of the major surprises of the projects described in *Barriers and Bridges* was the discovery that large ecological and social systems are almost by definition more dynamic and unpredictable than previously thought (fig. 5.8). The case studies illustrate profound and potentially "transient" pathologies of resource management. As with the pastoralists and fishermen in this book, innovation emerges when existing approaches are clearly inadequate and new approaches need to be designed. The authors emphasize developing a coherent theory of adaptive approaches to complexity and learning to harness change, rather than fight it. A challenge addressed in a science of open spaces is recognizing that effective science and other forms of knowledge gathering must be intimately bound to the underlying social fabric, while being both strategic and opportunistic.

"In the Weeds"

Despite the optimism reflected in *Barriers and Bridges* and *Compass and Gyroscope* before it, adaptive management has rarely been successfully applied in large, complex systems. Former University of British Columbia group member William Clark wrote in a 2002 editorial that the process "has yet to fulfill its promise in practice." Clark cited the limitations of institutional designs involved in large-scale socioecological "experiments" and recognized many challenges to effective implementation, "not the least the complexity of the linked ecological and social systems that adaptive management seeks to address and the high political stakes involved in the outcomes it seeks to influence."

Glen Canyon, part of the Colorado River system in the southwestern United States, is commonly cited as a pioneering example of the application of large-scale adaptive management, and as such it highlights the complexities involved in implementing large-scale environmental projects (fig. 5.9). Despite an initial emphasis on the design of inclusive decision

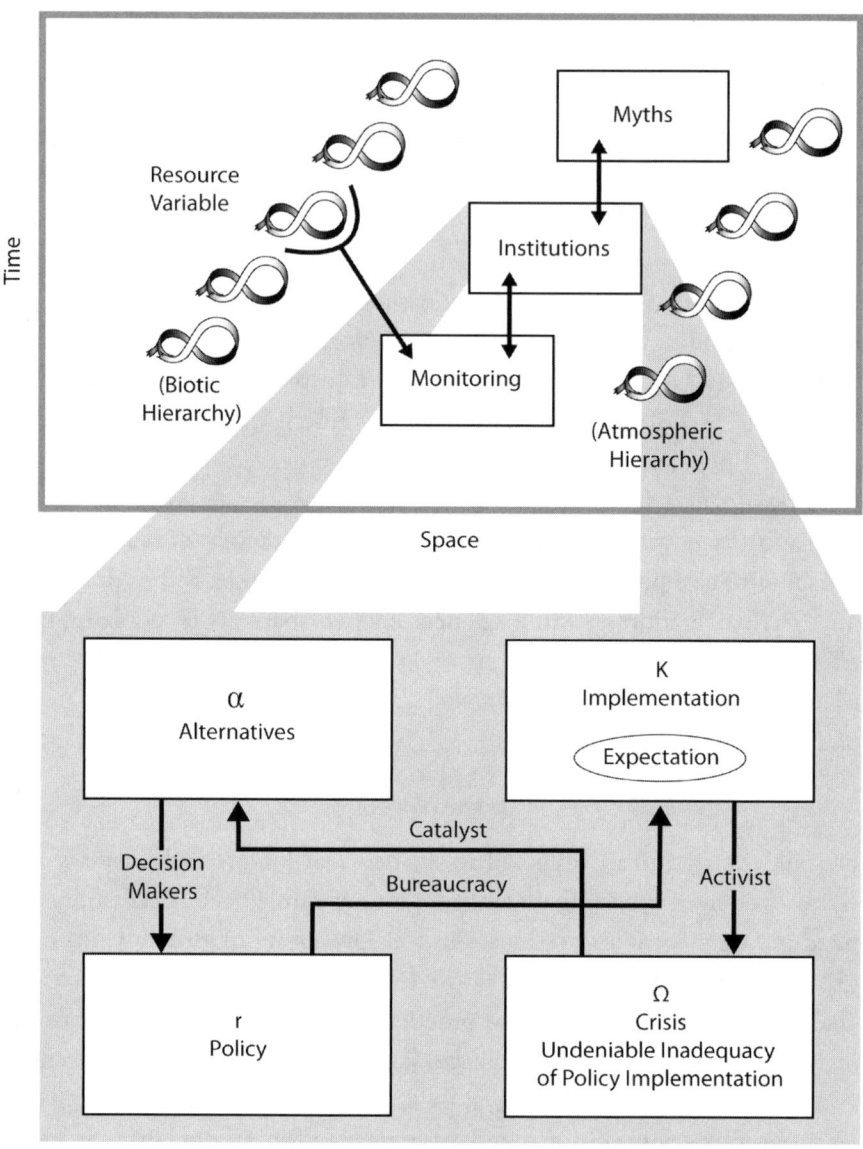

Figure 5.8. *The linkages between resource dynamics and variation in policy across time and space. The graphic illustrates the cross-scale attributes of social systems and policy and that although the adaptive cycle was founded from ecological theory, it also maps well onto social processes. (After Gunderson et al., 1995.)*

Glen Canyon Adaptive Management

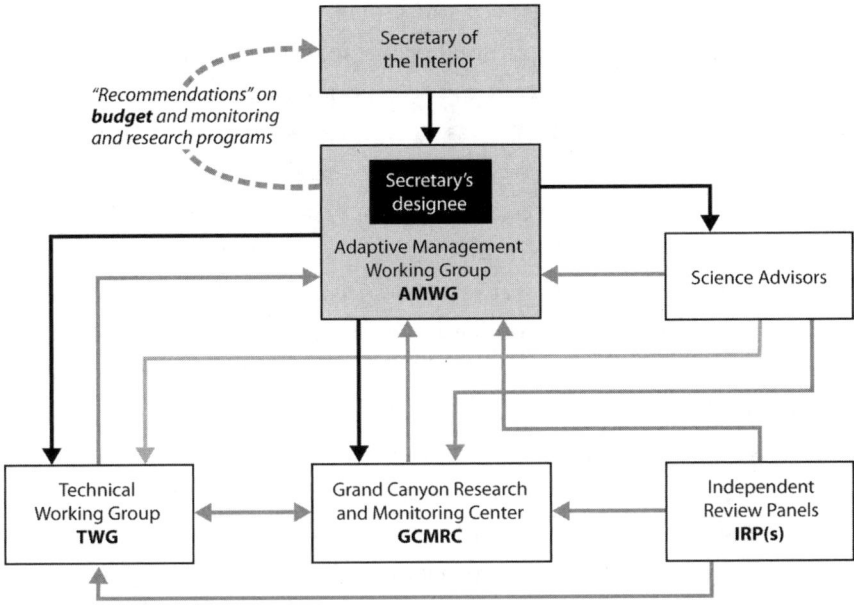

Figure 5.9. The design of the Glen Canyon restoration program with interaction among different levels of governance. Although a distributive process of governance was in place, power interests still managed to circumvent the process to maintain the status quo, which demonstrates the importance of governance design for equitable process and that even with the best of intentions plans can go awry. In this case the extensive feedback loops in place were not enough to prevent powerful interests in the Adaptive Management Working Group from controlling the outcome of the process (After David Mattson, unpublished.)

making, the science and policy-making process came to be dominated by technological fixes that largely excluded core stakeholder groups, such as the local tribes.[27]

At a cost of millions of dollars, various Glen Canyon programs addressed the surface outcomes, rather than the root causes of degradation in the Colorado River ecosystem, illustrating a classic type 1 approach to resilience embedded in a single-loop learning process.[28] For example, one restoration program spent hundreds of thousands of dollars mechanically removing trout to protect endangered endemic fish from being devoured. Since the construction of the dam, river temperatures were too low and the water not turbid enough for the native fish adapted to the

river's historic condition. In a Band-Aid approach that addressed the outcome, but not the root causes of the problem, the mechanical approach
satisfied the minimum requirements for action under the Endangered
Species Act of reducing trout, but it was not clear if a stabilization in
the numbers of endemic fish was due to trout removal or a period of
warmer, siltier river conditions.[29] Likewise, sandbars used for recreation,
which had been lost to erosion from changes in the river's flow due to
the dam, were restored by dumping sand in selected areas, rather than
by restoring the river's natural processes that led to their creation. Both
approaches are politically expedient, yet unsustainable. Endangered fish
continue to be threatened by environmental factors unaddressed by the
removal of trout, and newly created sandbars continue to erode without
deeper consideration of how rivers function. There are no shortcuts for
developing effective stewardship processes; design must encompass the
social as well as ecological facets of open spaces.

The results of the Glen Canyon experiment reveal the importance of
effective institutional design and show that modeling and science expen-
ditures often have only limited impact on policy decisions. According to
the U.S. Geological Survey scientists involved, deficiencies in the process
itself, as well as personnel interactions, decided the outcome, despite a
large investment in the project's science at the outset. That outcome was
largely to maintain the status quo, supporting the interests of electric
power producers over environmental, social, and spiritual concerns. The
process resulted in a compromise that does not address the heart of the
problem, which is how to fundamentally change the postdam construction flow of the river to better mimic historic flood cycles.

Another prominent example of adaptive management is the Florida
Everglades, which, despite millions of dollars spent on restoration and
protection, continue to decline. Water management strategies for conservation targets such as the endangered Cape Sable seaside sparrow have
yet to produce desired habitat conditions and may actually be negatively
affecting another rare bird, the snail kite, as the mangrove islands they
depend on continue a multidecade decline. A 2008 National Research
Council progress report concluded that the Comprehensive Everglades
Restoration Plan (CERP) "is bogged down in budgeting, planning, and
procedural matters and is making only scant progress toward achieving

goals, meanwhile, the ecosystems that it is intended to save are in peril." The report continues: "(1) the condition of the Everglades ecosystem is declining; (2) the CERP is entangled in procedural matters involving federal approval of projects and lacks consistent infusions of financial support from the federal government; and (3) without rapid implementation of the projects with the greatest potential for Everglades restoration, the opportunity for meaningful restoration may be permanently lost" (p. 223). One insider who has worked on the Everglades restoration for years noted that the optimism that the system could be restored and sustained has not proved valid because the process has not delivered the promised outcome of restored ecological function. He stated, "If anything, CERP has become a water supply project for cities along the Gold Coast."[30]

The results of these ecosystem-level experiments expose the soft underbelly of adaptive management, illustrating its tendency to rely on technology-driven approaches and top-down command-control governance frameworks, while not establishing true collaborative practices and effective information feedbacks. Kai Lee referred to this in a recent conversation[31] as "adaptive management lite," in which, although agencies and organizations go through the motions, the essential feedback loops are missing and the system fails to learn. So this is essentially management without the capacity for adaptation. Lee noted that he no longer has the optimism reflected in his 1993 book. "We are in the weeds," he stated, recognizing that we are in a much more complex and nuanced reality than envisioned by the discipline's founders, with the stakes higher than ever that we get the process right.

Despite the shortcomings of most adaptive management, one promising model is that of the Platte River recovery program begun in Nebraska in 2007. On the Platte management is undertaken by a governance committee consisting of all the major stakeholders, including the U.S. Bureau of Reclamation, the U.S. Fish and Wildlife Service, and the states of Colorado, Nebraska, and Wyoming. Yet in contrast to other programs, the committee hired an independent firm and executive director without a vested interest to oversee the day-to-day process. Although it is early days, thus far the results have been encouraging, for they suggest that independent oversight and implementation may be circumventing many of the issues of power abuse and distortion of the process seen in earlier efforts.[32] This suggests that a key to successful implementation of programs is explicitly addressing issues of power up front and establishing governance that ensures effective external oversight.

Undercurrents and Gaps

As recognized by anthropologist Paul Nadasdy in a 2010 critique, "most resilience theory deals inadequately, if at all, with questions of power." Issues of control are always lurking just below the surface, in megaprojects such as Glen Canyon and the Everglades, but also in the smaller place-based examples that form the backbone of this book. The Malpai borderlands example illustrates the power dynamics inherent in large-scale landscape stewardship. Over time, the interests of the Animas Foundation displaced the science process, even when the foundation was formed with the explicit intent of promoting conservation science. This resulted in short-term interests of a few individuals displacing the long-term conservation needs of an entire region. As scientists working on the project, we were bewildered that the more successful the program became, the more pushback we received from the foundation.[33] Although science was an immense asset, it came to dominate the conservation process, providing researchers with power disproportionate to their standing in the community. Later MBG undertakings, such as the strategic plan, diffused the influence of the science.

Similarly, in fisheries there is often a struggle for control between fishermen and science/policy. In the Maine lobster fishery, these power tensions are harnessed through governance at the local level to sustain ecosystem processes, whereas in the regional groundfishery, discontinuities between the scale of marine ecology and socioeconomic factors have decoupled the whole system.

To be successful, sustainable conservation and science in open spaces must develop common ground among interested parties, level the playing field, and implement necessary feedbacks to test assumptions and inject new knowledge into the process. The Malpai Borderlands Group was able to do all of these things effectively early in their development. Prior to formation of the MBG, borderlands ranchers felt disempowered by their distance from centers of governance, like the local groundfish fishermen in Maine.[34] Locally based science gave both groups the leverage they needed to engage in the process and reoriented the power relationships back toward the local community.

Though the issue of power may in hindsight seem so obvious as to be trivial, it is very much a reflection of the different institutional cultures and intellectual roots of disciplines. Ecologists do not explicitly consider power relationships any more than most political scientists explicitly consider biodiversity. Disciplines have different prisms through which they view the world, and only through transdisciplinary approaches can these very different perspectives be integrated.

The issue of power was brought home to me during a talk on the Malpai I gave at the community-based Collaborative Research Consortium back in 2005, during which a southern union organizer stood up and asked me, in a thick Mississippi accent, "Who has the power?" I had just spent forty-five minutes skirting around the issue, and she drove right to the point. In building sustainability and effectively conserving open spaces, addressing power relationships is not an issue—it is *the* issue.[35]

The borderlands are a microcosm of these complex power relationships situated at the intersection of the super-rich and hard-scrabble generational ranchers, agencies, and nongovernmental organizations. Independent scientists were the least powerful member group within this complex interplay, with only science-based knowledge as a lever by which to counteract powerful social dynamics that too often serve to direct conservation efforts toward short-term interests, rather than long-term sustainability. Which direction is ultimately embraced depends on the motives and social climate of the time. A science of open spaces is a fundamental departure from resilience science in that it is grounded within local contexts where issues of power and equity are recognized and integrated into conservation. Only by addressing these issues through science, governance, and policy design explicitly in an open and transparent process from the outset do conservation efforts have any hope of being viable for the long run.

Ecological Policy Design

The preceding discussion in this chapter illustrates that though transformational in their scope and implications, neither resilience nor adaptive management by itself can tackle larger and more pervasive environmental challenges. This is a drawback recognized from their inception, not just for the lack of consideration of social dynamics, but also for the fundamental constraints of an adaptive approach. Walters and Hilborn in the 1970s recognized use of the pesticide DDT as an example of these shortcomings, because it accumulated in food chains too slowly to be documented by predominantly short-term monitoring until it reached levels that severely affected the ecosystem. This illustrates the need not just for adaptation, but for governance and policy design to establish the proactive institutional frameworks necessary to develop and sustain the relevant science needed to address large, complex environmental problems.

A final dimension to resilience science—ecological policy design—sought to address many of the governance dimensions of adaptive management. Though it never gained traction after initial papers in the 1970s, it did reemerge in the 1990s under the guise of "sustainability studies" and as such the paradigm continues to be inflential.[36]

Despite representing the road not taken, ecological policy design was on to something important by recognizing the distinction between design and planning. It is rarely recognized that the larger implications of resilience science and adaptive management both spell death to conventional planning. But if one is to "expect the unexpected" and view management as a series of adaptive experiments, then the old strategy of building future predictions on past experience is impossible in our rapidly changing world. Using a series of overarching principles to adapt is not planning at all; it is design. This may seem like semantics, but the difference between design and planning is fundamental. The words are often used interchangeably, but their outcomes and processes are profoundly different.[37] Planning is "the act of formulating a program for a definite course of action," while design is "devising for a specific function or end."[38] Thus, design is a fundamental departure from conventional planning or management. Adaptive management is the process of adaptation to change, whereas ecological policy design sets in place the preconditions for anticipatory action. Devising effective science and policy lies in not just learning from the past or adapting to the present, but designing for the future.

For these reasons, I have proposed[39] the term *resilience design,* which embodies types 2 and 3 approaches to resilience. This is an approach we will return to in more detail in the next chapter. It builds on the powerful aspects of resilience in anticipating dynamics, recognizing thresholds, and expecting change.

Resilience design, though it embodies elements of resilience and adaptive management, is notably different from conventional management and planning constructs in its anticipatory approach. A useful analogy might be made to skeet shooting clay pigeons with a shotgun (fig. 5.10). If shooters aim to hit the target where it currently is, they will always miss because by the time the shot pellets reach that location, the target will have moved to a new position. Successful shooters anticipate roughly where the target will be, so the clay pigeon catches up to where they aimed.

In our analogy, the cloud of shot from the gun represents a range of possible solutions to a conservation challenge, rather than a single, fixed, precisely predetermined outcome that is almost inevitably wrong. Just

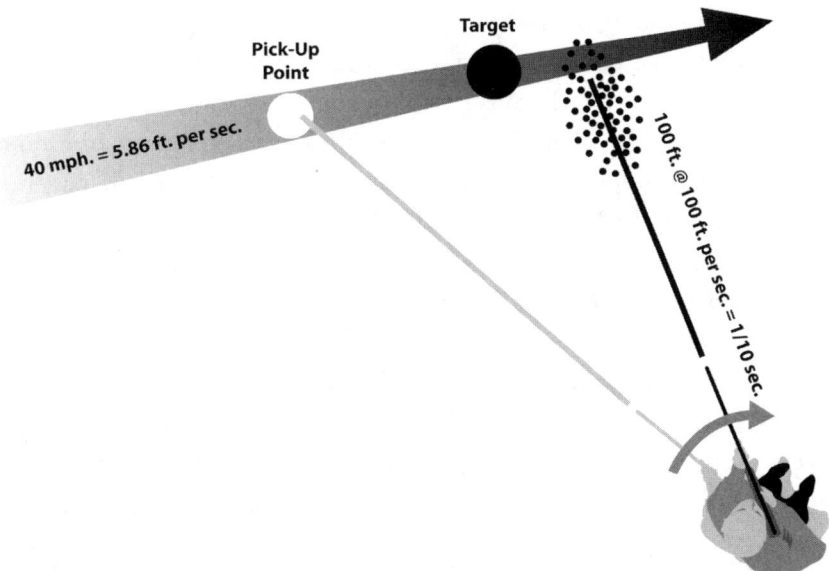

Figure 5.10. *The complexities of policy design are represented in this simple diagram of target shooting with a shotgun (skeet or sporting clays). The target is moving at one speed, the shot at another. To hit the target the shooter cannot shoot where the target is, but where it will be when the shot hits it. Effective approaches are proactive, rather than reactive, with the shooter not only anticipating where the target will be, but also the shifting optical illusions the target presents at different points in its flight path.*

as good shooters realize that their perception of the target is an optical illusion that changes depending on angle, distance, and myriad other factors, practicing resilience design involves recognizing similar "illusions" and adjusting one's approach.

Concluding Remarks

Through the merging of three broad and related streams of thought (resilience, adaptive management, and ecological policy design), resilience science leads to an international paradigm built upon the realization that change is inevitable and that science and management must approach the world based on that assumption, rather than one of stability. Resilience science treats actions as experiments to be learned from, rather than intellectual propositions to be defended or mistakes to be ignored. In a world filled with complexity and uncertainty, surprise is inevitable, so resilience science seeks safe-to-fail options that promote experimentation and learning, while reducing the risk of catastrophic loss.

However, in practice, most resilience science largely represents, at best, type 2 solutions, because the approach is too often reactive, rather than proactive. As Glen Canyon and other examples show, these bold experiments, though innovative and dedicated to testing management ideas, are frequently compromised by institutional structures that essentially maintain the status quo because they do not address issues of power or establish truly collaborative process. Don Ludwig, one of resilience science's founding thinkers, boldly proclaimed in a 2001 article, "The era of management is over." Conceding that the complexities involved mean it is hubris to assume that people can control large complex systems, he went on to say, "There is ample evidence that systems approaches and management are inappropriate for the complex ('wicked') problems that are most important today." Ludwig characterized these problems as "radical uncertainty" and "plurality of legitimate perspectives," indicating that more syntheses and dynamic approaches are needed, as the fisheries, rangeland, and adaptive management examples demonstrate. A science of open spaces builds upon the lessons from resilience studies through recognition that adaptation and planning are often insufficient and that strategic, design-based approaches are essential for creating the institutional sideboards to allow for more effective science and policy.

Ludwig's commentary is a tacit reminder that conventional approaches to biological and social sciences, or even systems approaches as typified by resilience science, do not fully address large-scale conservation issues. Complex, post-normal paradigms as presented in a science of open spaces are key to sustaining large systems with a focus on design that integrates physical, social, and ecological perspectives in a place-based context to generate the adaptive capacity to take advantage of opportunities or respond to threats.

In the next chapter, we integrate lessons from the resilience paradigm, coupled with the preceding discussion of theory and practice, to review strategies for implementing a science of open spaces. We need grounding in both conceptual and pragmatic perspectives in order to craft durable and effective science and policy to address the existing challenges of sustaining large complex systems.

CHAPTER SIX

Practical Aspects of Sustaining Open Spaces

Alice: Would you tell me, please, which way I ought to go from here?
Cheshire Cat: That depends a good deal on where you want to get to.
Alice: I don't much care where.
Cat: Then it doesn't much matter which way you go.
Alice: So long as I get somewhere.
Cat: Oh, you're sure to do that, if only you walk long enough.
        —*Lewis Carroll,* Alice's Adventures in Wonderland

In this final chapter, we focus on the practical aspects of conserving open spaces. Conventional approaches to science and policy often fall short in their relevance to the large and complex issues facing society because small and short-term or single-discipline studies are not sufficient to address the challenges typified by such issues as climate change and habitat fragmentation. Rather than applying more regulation and management to control each aspect of the process, the perspective promoted here focused on designing the preconditions for long-term success. The key step is development of the institutional capacity to maintain and enhance coupled ecological and social processes. Building on the foundation of theory and practical experience discussed in earlier sections, this chapter develops a framework for on-the-ground action and provides basic guidelines for undertaking conservation within open spaces.

Addressing the Conservation Paradox

Conservation faces a fundamental paradox: relatively small-scale reserves are limited by biological constraints, large-scale landscapes by social and economic ones. With small areas, one can close the gate or erect a fence and intensively manage human-related disturbance. However, no matter how many resources are brought to bear, it will always be constrained by the "tyranny of space."[1] It is ironic that for the most part any natural area that is small enough to be easily managed will likely never be large or whole enough to be ecologically sustainable.[2]

The opposite is the case in large landscapes, which are not limited by space, but by process. The irony here is that the larger and wilder a system, with the possible exception of the most remote and uninhabitable spots on the planet, the more sophisticated the science and policy that is needed to sustain them. For people cannot just manage their way out of situations in a large and complex system, but must harness the very force that most threatens the system's survival: people themselves. This requires approaches that are proactive and strategic, especially in the face of uncertainty.

Shooting the Rapids

Practitioners and theorists use the analogy of "shooting the rapids" (fig. 6.1) to describe the process of designing and implementing conservation practices or managing resources.[3] In navigating whitewater rapids, preparing for the initial descent is crucial. Paddlers first scout a navigable route from the bank, because once they are in the river and the current takes hold, opportunities for course corrections are usually limited.

In rapids, as in conservation, some situations are relatively straightforward; you can see the path ahead, anticipate obstacles, and plan accordingly. Rittel and Webber[4] referred to such challenges as "tame problems." Akin to class I or II rapids, they are easily navigable and the hazards are well defined. If questions over direction arise, you can always pull out of the current to scout the next steps at clear stopping points along the route. Much of the field of planning assumes that the world works this way. Unfortunately, reality is rarely so straightforward.

Figure 6.1. *In whitewater paddling, as in environmental science and policy, scouting the route and planning a path forward is key for success, with the larger and more complex the system, the more strategy and design needed. (Photo courtesy of Shutterfly.)*

Rittel and Webber used the term "wicked" to describe more complex problems, analogous to class IV or V rapids. For these, there are no simple and straightforward solutions, just better or worse outcomes for optimal situations are not an option. Such rapids are extremely turbulent, providing little opportunity for pausing once you are in the flow. There is no simple formula for getting through; you must choose the best path and improvise when faced with challenging situations or adapt when presented with new information.

Increasing rates of social and ecological change are creating higher levels of turbulence and greater challenges for conservation science and management to navigate. Unfortunately, technological advances, from the internal combustion engine to the Internet, tend to further exacerbate the situation by increasing the short-term illusion of control, at the cost of longer term sustainability and increased unpredictability.[5] Effectively addressing "wicked" problems requires understanding their underlying processes and pathologies to isolate potential solutions from an almost infinite number of traps.

So how does one navigate the safest path down the river in such turbulent times? Systems theorist Donella Meadows spoke of twelve policy "leverage points" that facilitate effective action.[6] These leverage points, for our purposes, can be distilled down to eight principles essential to the practitioner's tool kit. These are the basic concepts necessary for creating durable approaches to policy and maintaining open spaces. They are presented in order of highest to lowest impact, recognizing that many of these options may not be available in any given situation.

1. The influence of paradigms and values. Society has shared ideals—unstated assumptions that drive much of its legal and procedural actions. They are our shared sense of what is reasonable and fair—societal conventions that are deeply held, but not wholly unassailable. Though these social norms or personal values are difficult to change, shifting them can be hugely important in setting the stage for all that follows.[7] The circumstances leading to the formation of the Malpai Borderlands Group (MBG) provide a perfect example of shared perspective that needed to be overcome, namely, the suspicion and enmity toward conservation agencies and organizations long held by area ranchers, and the similar assumptions held by conservationists. Challenging these underlying assumptions can lead to new paradigms, but—as seen with the MBG—considerable effort is often needed to sustain these approaches over time.

2. Goals of the process. Goals determine outcomes; thus, being clear about the intended outcomes of a process is key to its ultimate effectiveness. In the Malpai science program, having a joint vision of the challenges recognized by the ranchers and researchers united the group's purpose. By contrast, in fisheries conservation in Maine, it was much rarer to find a united focus, and without a unifying view of the questions to be addressed, it was all but impossible to develop consensus. For as with blind men describing an elephant, one group may describe the problem as overfishing and another as climate change, while still others see the need to restore forage fish. The parties talk past one another as an outcome of their radically differently perceptions of the problem. In conservation, the old adage is especially true that if you do not know where you are going, you will never get there, or at least you will be walking a very long way, as Alice found in her conversation with the Cheshire cat at the outset of this chapter.[8]

3. Power of self-organization. Sustainable processes are, almost by definition, self-organized. For example, the actions of the Maine lobster fishery have largely been effective because they were not planned or directed from the top down, but emerged once the right social preconditions were established and governance sideboards maintained. The same was true in the Malpai borderlands with respect to fire management, which ended up a powerful integrator of other system properties, such as the amount of ex-urban development or the extent of contiguous unfragmented rangeland. Effective assembly rules are built around those parts of the system where small shifts in initial conditions cause profound changes in outcome. The Massachusetts Institute of Technology's Peter Senge[9] used the example of ships as fulcrums; it is impossible to change a ship's direction from the bow without huge amounts of force that will likely cause damage. However, from the stern it takes only a small shift of the rudder to alter the ship's course. Assembly rules are like a rudder, steering science and policy from the point of optimal leverage and transforming the direction of discourse and action.

4. Rules of the road. The legal, administrative, and social rules of a system—essentially its incentive structure—profoundly influence conservation outcomes. For example, among the Maasai, the expectation of reciprocity among family members, even distant relatives, sustains a scale of social and ecological function that generates large-scale patterns of landscape-level diversity by reducing the damaging impacts of maintaining cattle in the same relatively small area during drought. Conversely, in the case of ground fisheries, the New England Fishery Management Council process led to decisions driven largely by politics, whereas in the case of lobster fishing, the preconditions for restoration and sustainability through local control were put in place through a few simple rules that promoted local engagement and stewardship. As Meadows noted, "If you want to understand the deepest malfunctions of systems, pay attention to the rules, and who has power over them." Power issues played a big role in the failure of large ecosystem programs such as Glen Canyon and the Everglades because economic interests trumped ecological concerns. The successes and failures of all of the case studies can be directly tied back to the rule sets and the power structures they represent.

5. Structure of information flow. A crucial insight from the MBG example was that the coupling of science-based and local knowledge created the credibility to undertake activities such as land conservation and the resources to influence federal agencies to undertake landscape-level reintroduction of fire. The Maasai pastoralists and the Maine lobstermen displayed a more informal, but equally tight, coupling between information and action. The Maasai in their daily milking and the lobstermen in their daily catch each provide a regular assessment of environmental conditions to which the resource users can readily respond.

6. Impact of feedback loops. Negative and positive feedback loops have fundamentally different outcomes. A positive feedback loop is self-reinforcing; a negative feedback loop is self-correcting. Positive "success for the successful" loops compound the impacts but also provide less control. Negative feedback loops, by contrast, typically include a specific goal or braking function.

Fisheries use both positive and negative feedback-loop approaches. For example, in the lobster fishery, fishermen follow local rule sets in a positive feedback approach that emerges from local norms and regional consensus. The groundfishery uses a negative feedback-loop approach with specific set points, such as "days at sea," designed to regulate catch levels. Meadows noted that negative feedback approaches tend to be extremely resource intensive to implement. In large landscapes, due to their size and complexity, it is almost essential to use positive self-reinforcing feedback loops in governance and to avoid negative feedback approaches whenever possible.

7. Influence of time lags relative to the rate of the system. A system can not respond to rapid, short-term changes through slow, long-term processes. It is possible to slow rapid changes to match the system's capacity for adjustment, but usually it is more effective to alter the scale of focus of the response, rather than try to manage the change itself. For example, in finding consensus over climate change, reaching agreement at the international and national level has proved difficult, although numerous small municipalities have been remarkably adept at rapidly instituting local policies.[10] Likewise, the processes of collective impact (discussed in chapter 3) uses backbone organizations to network numerous local organizations, rather than relying on highly centralized international agreements or overarching top-down governance schemes.

8. *Size of buffers relative to their flows.* Buffers stabilize complex systems by dampening short-term variability. Imagine if liquor stores had to order from a brewery every time a customer purchased a beer. Instead, keeping an inventory allows the merchant to respond to short-term demand. Although expanding buffers (e.g., inventory) can stabilize systems, this can be expensive, and having too much investment in inventory has its own consequences. To continue with the beer metaphor, a store would be foolish to precisely predetermine which kinds of beer it will stock for an entire year because consumer demand may change. Instead, stores adaptively manage, letting signals from the market guide their short-term inventory decisions. The need for substantial buffers is why the ground-fishery management approach does not work; maintaining the necessary buffers to avoid compromising populations during periods of low recruitment is rarely politically expedient because of the economic hardship it imposes.[11]

Meadows's insights have proven to be immensely influential in applying systems principles to real-life situations, and yet the devil is the details. Less clear is how to translate the theory into practice. One possible approach is through what I call *resilience design*, which builds on the lessons of resilience and adaptive management and uses a handful of practices and ground rules to alter overall system behavior. Its three core elements are (1) *equity design*—collaborative approaches that generate effective self-organization; (2) *process design*—developing effective knowledge through the establishment of more relevant and appropriately scaled science and monitoring; and (3) *outcome design*—a combination of the previous two. Outcome design is primarily concerned with the practical aspects of sustaining organizations long enough and at scales large enough to make a significant difference in conservation outcomes.[12]

Resilience design applies post-normal approaches to large-scale conservation science. This includes moving beyond resilience science's base in academia and agencies to apply the principles of a science of open spaces through a place-based approach that embraces local social dynamics and accounts for power dynamics in conservation design.[13]

Equity Design

Solving complex problems requires breaking down barriers and establishing connectivity.[14] Shared cognition is essential for success, as demonstrated by the MBG, which broke down the divides among ranchers, conservationists, and researchers to create synergy for effective land conservation.

The properties of personal and group collaboration, as illustrated by the cross-site coordination among pastoralists such as the Maasai in southern Kenya (discussed in chapter 1), demonstrate the power of collective thought and action and why these approaches, although complex and time-consuming, are more sustainable than command-control governance that tends to be brittle and unresponsive to change. As argued in previous chapters, the world is intrinsically dynamic, and governance arrangements need to be inherently flexible and adaptive while having a clear sense of their goals and expectations.[15] The foundational principles from physical, biological, and social perspectives in combination illustrate that we need institutions that generate appropriately scaled feedback loops and interconnections for adaptation through learning. For this reason, double- and triple-loop learning approaches are essential for responding to change, but are also anathema for command-control governance that thrives on short-term control, accumulation of power, and the illusion of stability.[16]

From organizational theory developed by researchers such as Argyris and Schön to the practical experience of practitioners, successful action ultimately depends on effective governance principles that engage people in an equitable, open, and transparent process.[17] Without collective goals and common ground, people experience social fragmentation, resulting in competition and conflict.[18] Effective governance for science and conservation balances short- and long-term interests and mitigates power imbalances. Effective governance can be both implicit by using a few informal assembly rules to develop emergent outcomes that sustain self-organization, or explicit through developing clear goals and overarching guiding principles.[19] For example, the Malpai Borderlands Group used an approach to fire management that coordinated diverse fire management organizations, while having relatively clear, but flexible, expectations about desired future conditions of the landscape set by each landowner.

Social and ecological fragmentation are closely aligned, for both are outcomes of breakdowns in connectivity. Among the numerous ways of rebuilding connectivity, Herman Karl, of the U.S. Geological Survey, and colleagues from the Massachusetts Institute of Technology developed a process called joint fact finding (fig. 6.2), which represents a synthesis of best practices in natural resource decision making.[20]

In joint fact finding, although a mediator can help organize interactions among disparate interests, finding consensus does not need to be mediation-driven. Instead, the intersection of science and local knowledge can play a key role in "community building" by generating common ground and adaptive capacity to address complex information needs in a more deliberative process. To be viable, solutions cannot be just political compromises, but must find common ground within the context of the realities of the resource.[21]

The Malpai example demonstrates that community building can be undertaken directly by individuals or organizational leadership, or indirectly through joint knowledge generation, as in the case of collaborative science, such as the McKinney Flats project. In the borderlands, The Nature Conservancy originally played the role of convener, bringing diverse groups together. The Nature Conservancy's logistical expertise and access to funders and capital were key to making the vision of the ranchers a reality. The borderlands example also illustrates why a balance of leadership between locals and outsiders is all but essential. In the early years, The Nature Conservancy shared leadership responsibilities with local ranchers. Not only was the symbolism of "big hats and baseball caps" (i.e., ranchers working with conservationists and researchers) especially effective at attracting funders and a powerful symbol at congressional hearings, but it also effectively balanced and leveraged the strengths of each group.

One major reason why a place-based collaborative should keep a portion of its leadership external is that in a close-knit rural community or any tightly linked group, there are limits to what one neighbor can ask

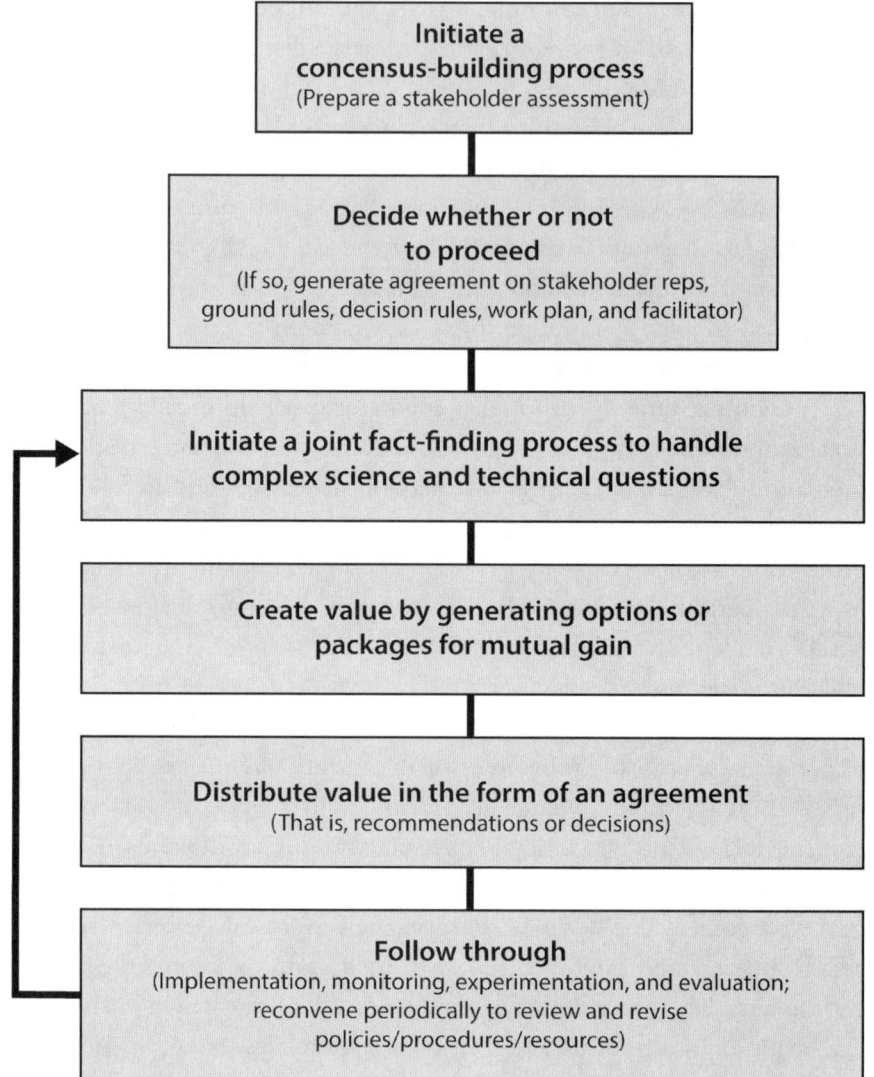

Figure 6.2. *Joint fact finding, by which science is developed with partners, lays out guidelines for collaborative science. The approach not only clearly spells out guidelines for success, but also shows how it is critical to engage partners at the outset of the process to promote trust and buy-in to the program (After Karl et al., 2007).*

of another, whereas an external person has more latitude to drive the process and keep competing interests in line.[22] In the MBG example, the approach to science changed almost immediately following TNC's departure as coleader. The rancher who assumed the role of executive director was as committed to science as TNC was, but as a longtime member of the local community, he needed be more careful to not compromise his relationships with friends and neighbors. The change from joint to solely local leadership coincided with reduced scientific input into the policy process due in large measure to these social constraints.

The Malpai case study in chapter 2 illustrates how increasing distance between the scientists and the policy process sowed the seeds for the eventual decline in researchers' ability to inform the process. Scientists also provided expert advice that was less influenced by the political aspirations of associated agencies or organizations, following eminent ecologist James H. Brown's call for scientists to provide the group with "what they need to hear, not what they want to hear." This kind of pragmatic, critical assessment was lost as the process became politicized and advisors increasingly curried favor with the ranchers by playing to their sensibilities, rather than taking tough, if sometimes unpopular, positions. The reduction in objective input was one of the greatest losses to the conservation program, for it led to a cycle of decline in effectiveness of both the science and the collaborative process.[23] Equity design, therefore, requires collaborative approaches that are sustained through carefully crafted, governance-developing, institutionalized processes of critical self-reflection, with periodic external review to avoid the pitfalls of groupthink and prevent short-term demands from trumping longer term goals and objectives. The point is to balance power relationships and make sure that critical review and the testing of assumptions remain core parts of the process.

Equity design also involves a dynamic balance analogous to Kai Lee's metaphor of the compass and gyroscope discussed in chapter 5, which describes the tension between science and politics in the adaptive management arena. The compass and gyroscope operate at the macroscale, examining the interaction of institutions of adaptive management and politics in large ecosystem projects such as the Columbia River. However, similar dynamics also operate at the microlevel within communities, where a dynamic tension exists between stabilizing forces of local interests and the destabilizing forces of science, new knowledge, and associated change. This dynamic tension is much like the forces of "understeer" and "oversteer" in the handling of a speeding car (fig. 6.3). Understeer makes the vehicle stable, but also unresponsive and slow to change direction. Conversely, oversteer, in which the vehicle is prone to turn too sharply, makes the handling unstable, because the vehicle has a proclivity to swap ends and spin out or lose control. At high speeds or with sharp cornering, a skilled driver seeks a dynamic balance between opposing understeer and oversteer, because each, in turn, influences the car's handling, just as effective policy skillfully applied also sustains a similar dynamic tension between stabilizing and destabilizing forces.

Equity design, therefore, requires instituting a continual learning process that can balance the metaphoric understeer and oversteer, building the institutional capacity to balance stabilizing and destabilizing forces. For example, any inputs, such as science or critical review, can be destabilizing in the short term, but are essential for the long term. Sustaining the dynamic tension that maintains open spaces essentially boils down to five ground rules:

- Maintain an open and transparent process.
- Engage all stakeholders, or their representatives, in the process.
- Build a shared vision of the system and an understanding of the opportunities and constraints.
- Ground decisions in empirical documentation and/or consensus with a framework for self-reflection and periodic external review.
- Instill permanent yet adaptive processes that foster adaptation and innovation while maintaining continuous cycles of double- and, the potential for, triple-loop learning.

All policy and design for effective and sustainable large-scale science and governance emerges from these five essential design principles, stemming from a strong base of collaborative process to build and sustain social capital. The next step is to ensure that the feedback loops are in place that can inform and sustain these dynamic processes because a continual flow of information maintains the communities' ability to evolve and respond to change.

Figure 6.3. *High-speed cornering in a car illustrates the dynamic tension between understeer, in which the vehicle plows, and oversteer, in which the handling is twitchy and the vehicle prone to swap ends. In the course of even a single turn these opposing dynamics are transient and prone to change based on weight loading, acceleration, deceleration, tightness of the corner, and numerous other factors. This example is a metaphor for the complexities of developing policy that promotes stability, and yet is nimble and responsive to change. (Photo courtesy of Shutterfly.)*

Process Design

Process design depends on working across scales of resolution to sort through the options and find those that create the most leverage for sustaining open spaces. These points of leverage lie at the interface of ecological, economic, and social systems where both the risks and the opportunities are greatest.[24] Finding these fulcrums requires integrating knowledge types in which the world is viewed through different paradigms and at different scales (fig. 6.4). Refining this knowledge and formally communicating it help to build the intellectual capital upon which scientific progress, and social durability, are grounded. The core tools to accomplish this include setting the context by attaining practical knowledge, understanding the scale of processes and extent of change through monitoring, and refining knowledge and testing assumptions through experimental science.

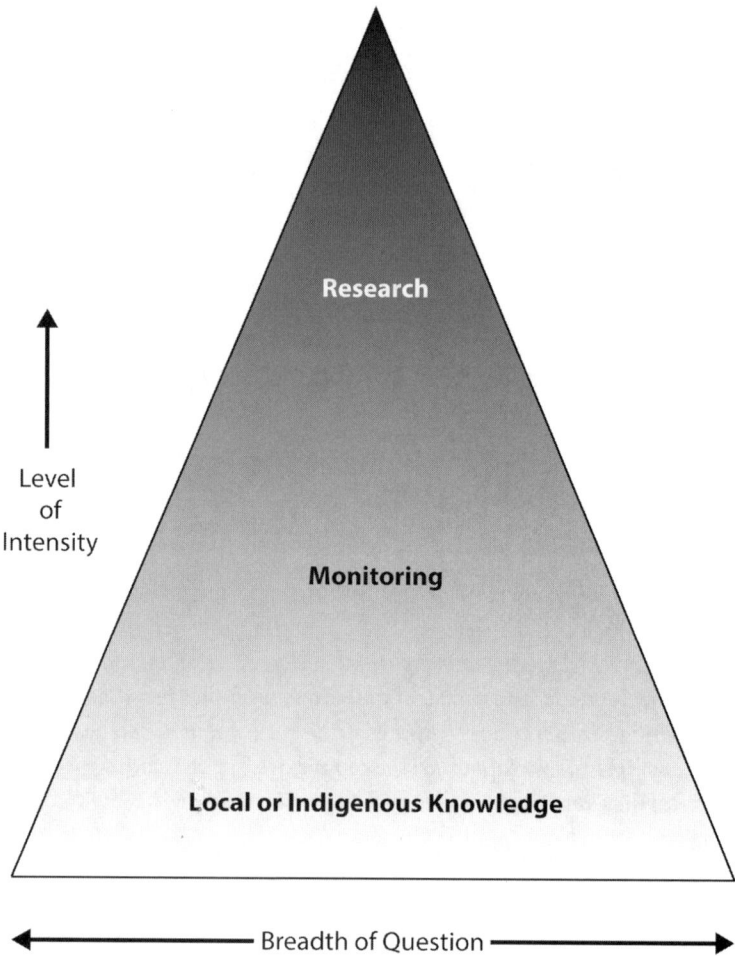

Figure 6.4. *In knowledge gathering there is an inverse relationship between intensity and breadth, with trade-offs between broad and low-intensity approaches, such as local knowledge, and intensive but more narrowly focused experimental research. These trade-offs between intensity and breadth suggest the necessity of integrating multiple knowledge types to understand complex systems such as open spaces.*

Practical Knowledge

The more complex or "wicked" a problem, the more informal the means needed to address it. System theorist Peter Checkland made a distinction between "hard" and "soft" systems.[25] Hard systems exist where formal experimentation and empirical data are possible. The classic example of a hard system is the 1960s NASA space program, whereby a few equations calculated on the relatively simple computers of the period, could send a rocket to the moon. In contrast, soft systems are often social arenas where quantitative prediction is considerably more difficult due to the sheer complexity of interactions and usually requires a more descriptive approach.[26] Large-scale conservation is intrinsically a soft system with a vast array of interacting variables, ranging from climate to land tenure.

The foundation for navigating soft systems starts with local and traditional ecological knowledge, the cultural capital by which societies convert natural resources into economic goods and services and sustain their populations. As illustrated in the rangeland and marine fisheries examples, the power of local knowledge is its ability to bring diverse groups together to broaden understanding.[27] The challenge is to find places of common ground where different perceptions can come together to generate common points of interest in the development of a joint vision of the overall system (fig. 6.2). Local knowledge is also valuable in its own right, for it provides an essential context for decision making by generating crucial questions that need to be addressed through more quantitative forms of information gathering—much as the borderlands programs arose from the rancher's questions and perceptions—while simultaneously addressing fundamental ecological theory.[28]

Monitoring

Monitoring is commonly defined as a means of tracking long-term change. But it also has a second, almost more important role of documenting the range of variation within a system and, as such, is key to establishing the appropriate scale of conservation, science, or stewardship. The level of design or scale of action must transcend that of ecological or social disturbances. All conservation and management is contextual, so understanding the role played by disturbance processes, such as fire in the West or hurricanes in New England, is crucial to designing for resilience. For example, core protected areas must be larger than the typical scale of disturbance so they are not completely eliminated by it (fig. 6.5). Conservation below the scale of disturbance frequently leads to destruction of ecological processes, whereas conservation above the scale of disturbance can lead to a mosaic of habitats that increases landscape and species-level

Figure 6.5. *The relationship between scale and disturbance suggests that an understanding of the scale of disturbance is directly related to the minimum scale of conservation. Conserved areas must exceed typical areas of disturbance to ensure that functional systems are sustained, because it is not a matter of if, but when, such an event will occur. A classic example is gap creation by windthrow in forests, which introduces diversity to the system. Some damage from a hurricane in New England illustrates how these meta-disturbances, though separated by decades, are a part of the system and how sustainable reserves must be larger than the extent of typical disturbance events. (Photo courtesy of Shutterfly.)*

diversity and ecological health. The role of large-landscape conservation is to transcend disturbances (both ecological and social) and incorporate them into the mosaic of different habitat types or institutional structures to maintain the adaptive capacity to respond to change or withstand surprise events.[29]

The need for monitoring can be seen in assessing the distribution of fire across the landscape, or in documenting vegetation change through time. However, monitoring must account for key driving variables. For example, in the Southwest, monitoring changes in vegetation is of little use without also documenting rainfall or grazing intensity. Otherwise, it's unclear whether declines in grasses resulted from drought or mismanagement. Without an environmental or land-use context, there is no means of interpreting the data. This illustrates that undertaking effective monitoring is relatively simple in theory, but extremely difficult in practice, as we found in the borderlands, where even the monitoring designed by a science committee containing many of the region's leading researchers still turned out to have shortcomings. Typically, it takes years of experience and critical evaluation to assemble a design that works under the range of conditions that affect a given area. Practitioners need to understand and account for the great deal of thought, planning, and trial and error required to get it right. The resource management literature and agency and conservation organization handbooks are replete with calls for monitoring to document system change. However, the process of monitoring itself is rarely evaluated.[30] It is not enough to just monitor, one needs to be certain that the data gathered actually address the most crucial factors in the most effective manner with adequate replicates that take into account associated driving variables (such as the aforementioned rainfall and grazing).

Monitoring is an important part of effective decision making. However, as commonly practiced, it all too often represents institutionalized single-loop learning in which people document what they expect to see, rather than learning from the data. So part of the design process should be to ensure that monitoring is not conducted for its own sake, but generates significant new insights. Monitoring is also often a political solution that provides the illusion of action without accompanying insight that might force activity or challenge existing paradigms, so it is essential that procedures exist to ensure that the information is not only relevant and effective, but also that it is actually used.

Experimental Science

At the highest level of research intensity are experimental approaches, studies that can be conducted only on relatively small portions of the landscape because of the relatively high investment in time and resources. Because of these constraints, experimental science must be used judiciously and is most effective when applied near ecological boundaries where system dynamics are most likely to be revealed.[31] Although experimental science is not free of observer bias, well-designed experimental studies are critical in allowing scientists and their collaborators to empirically test their assumptions. Perhaps more than any other level of inquiry, well-designed experiments, especially ones that extend over long time periods or across large spatial scales, can reveal unexpected results. The challenge is to develop science that yields fundamentally new insights—the triple-loop knowledge that is transformative and essential for effective response to change. However, there are two considerations.

First, more precision does not mean more accuracy. There are innumerable cases of highly precise studies that are ultimately irrelevant because they ask an irrelevant or uninformative question, or ask an important question at the wrong scale, yielding results that are misleading or limiting.[32] We have already considered examples from rangelands and fisheries that illustrate how scale influences the outcome of research. Considering the broader social context is profoundly important for developing more effective experimental design. For example, the Malpai Borderlands Group's cooperation facilitated a level of science not accessible through conventional approaches developed in isolation from the local community, and fishermen's insights about historical spawning grounds transformed the context within which fishery research was conducted.

Second, research is most effective when conducted at time frames sufficient to capture discernible patterns of environmental variation. It is better to document a few simple measures well over a long period of time than a broad range of variables poorly for a short period. This means that in large and complex systems, it is important to work back from what can be sustained long enough to reveal longer term patterns (ideally at least ten years), while building in periodic external review to reassess the process and ensure its continued relevance. Unfortunately, the grant system often requires that a project grow and evolve for its funding to be renewed. The need to tailor research priorities to meet shifting funding priorities too often generates short-term benefits at the expense of long-term continuity and good science, especially in large, complex systems that integrate social and ecological variables, and where scale-relevant research can take decades. This is where social design becomes essential, for to sustain the research program long enough and at scales large enough to be relevant to conservation and management entails the development of collaborative partnerships. The example of the McKinney Flats project in the borderlands illustrates both the benefits and pitfalls of undertaking the necessary large and dynamic, post-normal approaches.

In sum, when seeking leverage points and developing feedback loops, it is clear there are no shortcuts to gaining effective knowledge; a combination of approaches is usually needed. The intersection of local knowledge, monitoring, and experimental science is essentially triple-loop learning, developing a synthesis among the hard-won knowledge of local practitioners, the more formalized learning process of monitoring, and experimental science that tests underlying assumptions in a safe-to-fail environment.

Process design promotes interaction among diverse knowledge types in an integrated framework. This means that knowledge gathering should not be segregated into monitoring versus experimentation or social versus natural sciences, but developed as an integrated whole.[33] Six questions should be asked at the proposal stage of a research project:

- Is the knowledge gathered used?
- Is the knowledge gathered sustainable and cost-effective?
- Is the knowledge gathered integrated with other approaches to learning?
- Is there a process of critical external review and dissemination?
- Are the metrics leading or trailing indicators?
- Is there a process to integrate the results into governance and policy development?

A "no" answer to any of these should trigger reevaluation. These questions may seem obvious, but in my experience relatively few organizations ever formally ask them.[34]

To span the gulf between what academic writing describes and what practitioners face is ultimately where some of the most significant opportunities for promoting large-scale conservation occur.[35] However, to do so successfully usually requires an understanding of how to translate theory into practice and the establishment of an effective adaptive governance framework that encourages and facilitates both learning and knowledge transfer. Outcome design considers the practical aspects of facilitating durable and resilient means of conserving large, complex systems.

Outcome Design

Henry Mintzberg's classic article, "Crafting Strategy," (1987) in the *Harvard Business Review,* though devoted to understanding effective business approaches, has much to teach conservationists and policy designers. He stated, "In practice, or course, all strategy making walks on two feet, one deliberate, the other emergent." The point is to have a framework in place that maintains the state of the organization, while simultaneously taking advantage of opportunities when they arise. Conservation design needs to be "deliberately emergent" by developing processes to facilitate a more spontaneous, but reflective, response to change—to capture what Mintzberg called "discontinuities," those moments in time when being aggressively proactive or seeking change really counts.[36]

So how can such an approach be proactive without being destabilizing? A first step is to use a logic model that maps decision processes.[37] As with the "running the rapids" metaphor that began this chapter, a vision statement is essential for lining up the organization for the passage downriver. For example, the MBG's goals and principles, as stated in its vision statement, and widely used by other rangeland conservation organizations in the West, are

> to restore and maintain the natural processes that create and protect a healthy, unfragmented landscape to support a diverse, flourishing community of human, plant and animal life in our borderlands region. Together, we will accomplish this by working to encourage profitable ranching and other traditional livelihoods, which will sustain the open space nature of our land for generations to come.

MBG's philosophy was to be inclusive, open, and transparent based on the values of the community and standards of practice influenced by Quaker activist Jim Corbett.[38] This was accomplished by emphasizing peer-review-quality science and a standing invitation for critics to engage in dialog with the group. However, as the MBG case also illustrates, the drawback of relying strictly on a vision statement is that there is little guidance about how the vision will become practice, and how the practice will be sustained under changes in leadership. The MBG's vision statement is also a value statement, but it does not indicate how those values are attained or sustained, and completely avoids the issue of power relationships.

By contrast, the principle-based ethos of the Northwest Atlantic Marine Alliance provided an example for groups in the Gulf of Maine. Developed in 1995 by Visa credit card founder Dee Hock, fishermen, conservationists, and marine policy leaders, the group's message was simple: align diverse interests through basic principles.[39] It did so through a series of meetings to build consensus, not unlike those of the MBG of the same time period. Likewise, its vision statement was similar to that of the ranchers: "to restore and enhance an enduring northwest Atlantic marine system, which supports a healthy diversity and abundance of marine life and human uses."[40]

However, beyond this general mission statement, the Northwest Atlantic Marine Alliance also developed detailed principles of organization. For example, they sought to "vest authority in and make decisions at the most local level that includes all relevant and affected parties."[41] They also developed principles of practice, including "encourage adaptability, diversity, flexibility, learning, and innovation in all governance processes and practices." These instructions are considerably more detailed than those of the MBG, stressing the importance of developing preconditions for innovation that also sustain the integrity of the organization.

There are essentially two kinds of principles: ethical and procedural. For effective governance and resilience design, it is important to incorporate and follow both and to ensure that there are provisions requiring the principles be followed. What happens if the principles are violated? The implications of such actions are not trivial, as in the case of the Downeast Fisheries Partnership (considered in chapter 3). In this case, when the organization did not follow its own principles, some of the most senior and influential organizations crucial for the group's long-term success did not join the endeavor. In my experience, ethical breaches are the single greatest reason why conservation organizations do not reach their potential and why collaborative partnerships collapse. Too often, organizations assume that because they are doing important work, the ends justify the means. However, conservation at its core is an ethical venture, and ensuring that clear principles are maintained is essential to long-term success.

Especially as the group evolves through time, it is important to have clear guidelines about how the organization will conduct itself, as well as clear boundaries as to the group's priorities to prevent mission creep.

In the next section we transition from a review of levers for implementing effective or sustainable practice to their actual implementation. This is an area not often considered in the conservation literature, where it is assumed that the implementation phase is implicit. But it can often be the most challenging facet of the process.

Putting Principles into Practice

Effective practice is as much about pragmatism as idealism. In attaining durable solutions at evolutionarily relevant scales, one needs to develop the organizational infrastructure to sustain the process. How does one turn principles into practice?

Funding is the master key that links, locks, or unlocks much of the process of sustaining relevant programs. The marriage of finance and conservation is perhaps the ultimate driver of effective conservation and sustainable land use.[42] Funding introduces another scale issue (fig. 6.6). The problem with most grants and agreements is that they are generally short-term solutions to long-term problems. Some of the greatest threats and opportunities exist for sustaining open spaces in the context of this fundamental discontinuity.[43]

The greatest challenges reside in the transition period, from approximately seven to ten years, when a program has been around long enough to be old to funders but is not yet truly established. The MBG's science program is an excellent example. After eight years, it was well established, serving as an effective foundation for policy making, but having an increasingly difficult time acquiring the funding it needed to continue. By the tenth year, funding had all but dried up, despite MBG's well-regarded work. Rather than growing or evolving, the science program became increasingly simplified and conservative in response to dwindling levels of support. The irony was that the value of the research dramatically increased after the first years of the program even though the costs of experimental research dramatically declined. Ironically, by taking a short-term approach, funders often accrue the most expense for the least return from their investment.[44] The rarely taken long-term approach to

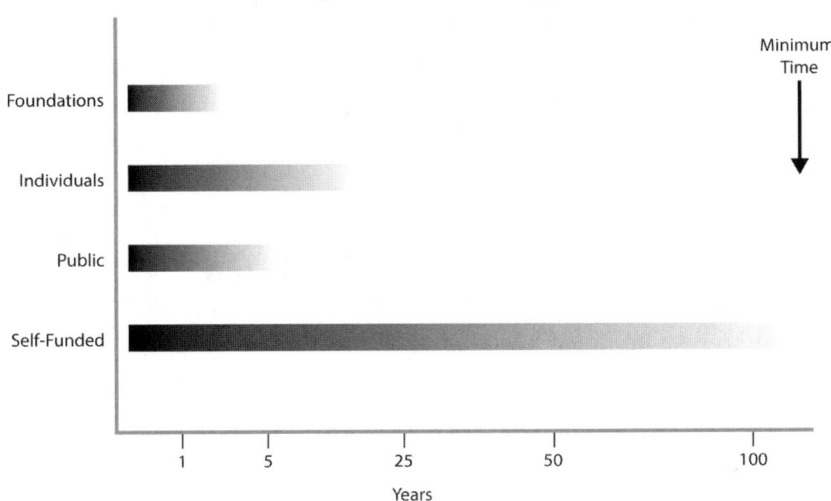

Pathologies of External Support

Minimum Time

Foundations

Individuals

Public

Self-Funded

| | | | | |
|1|5|25|50|100|

Years

Figure 6.6. *A fundamental disconnect in conservation is between the length of time a funder will remain engaged (typically three to five years) and the many decades it takes to attain large-scale progress. However, the academic literature largely ignores finance, which may be the most fundamental constraint on conservation. Developing strategies to promote longevity is essential for allowing conservation to operate at meaningful temporal and spatial scales.*

science and conservation investment, is what actually leads to the biggest payoffs.[45]

In the MBG example, the challenges seem to be largely institutional. Although the science had demonstrated its utility, these lessons were largely lost on the funding community.[46] In the end, the agencies, foundations, and even large conservation nongovernmental organizations that initially supported the MBG appear to have learned comparatively little from more than a decade of effort and over a million dollars of investment.[47]

Changing the Rules of the Game

The examples in the previous section illustrate how practitioners and the process of developing long-term, durable approaches have been dealt a losing hand under the current rules of the game. There is a fundamental disconnect between the scale of conservation, stewardship, and science and that of funding. One of the first steps to designing for resilience is recognizing some of the common traps and short-term illusions that thwart long-term progress. Developing an effective science of open spaces means developing the resilience design to transcend these traps. Some of the major pitfalls and essential responses are the following:

The Intellectual Trap

Disciplines—the silos that organize academia—often have little bearing in the "real world." There is considerable danger in applying strictly academic approaches, because complex problems typically do not fall into the neat and well-defined boundaries of intellectual disciplines. Though interest in socioecological systems has begun to change the emphasis of academia from single to transdisciplinary approaches, there are still relatively few rewards and incentives for cutting-edge research that spans disciplines.[48] Even the term *socioecological* is largely an artifact of an academic approach that seeks a synthesis, but still implies a dichotomy between disciplines.[49] By contrast, a science of open spaces seeks to break down this dichotomy and redefine the underlying incentive structure by demonstrating how transdisciplinary approaches grounded in place can expand the scale and scope of science and conservation.

The Private-Funding Trap

Grants and foundation support provide an illusion of organizational sustainability (fig. 6.6). Foundations, as with most organizations, need to grow and evolve, and thus they rarely stay with a given project for more than a few years, and frequently do not have a good exit strategy for when they leave. External funding is crucial for jump-starting a conservation program, but real sustainability comes from within and is the outcome of an effective design process that aligns ecological and economic constraints, goals, and scale. Therefore, external funding can be looked at only as a partial solution. Self-funding mechanisms that develop internal funding streams may be the only long-term means of ensuring a program's survival.[50]

The Federal-Funding Trap

Federal funding can be extremely political, often having fewer formal review guidelines than other types of support. But perhaps the bigger issue is the bureaucratic hoop-jumping involved with federal grants, which can be extremely time-intensive. I typically consider a dollar from a federal grant to be worth about half of one from private foundations due to the added administrative time and other costs. The Malpai example illustrated how there can also be extremely slow turnaround time in reimbursements, which can be crushing to smaller organizations. But more importantly, the relationships with federal funders need to be managed to keep abreast of rapidly changing political mandates or procedural changes that can severely derail the process. External oversight and other means of addressing the extreme power imbalance when developing partnerships with federal agencies should be built into the process from the beginning.

The Nonprofit Trap

The time and effort required to maintain a nongovernmental organization (NGO) are often underestimated. As with the rest of the federal government, the drive for accountability has increased the time and expense to nonprofit operations of meeting Internal Revenue Service (IRS) guidelines; small NGOs have been drawing increasing levels of scrutiny in recent years and becoming a popular target of IRS audits. Without the permanent legal and administrative staff or a means of readily funding such unpredictable costs, dealing with the inevitable entanglements with the IRS is an unanticipated challenge not built into most business plans. Even more problematic is that much of this tax work has become too complex to be handled by anyone other than an expert. Although many accountants claim to have expertise with NGOs, few actually do, and by the time the individual or organization discovers this, they are frequently on the hook for hefty fines and penalties or at a minimum a lot of unanticipated administrative work. One colleague who recently started an NGO went through three accountants before finding a reliable one, and she is an attorney with considerable tax savvy. NGO status should be a strategy of last resort, undertaken only after other options have been considered. For-profit status, or using a fiscal sponsor to handle grants and donations, may be much less expensive and time-consuming, allowing the organization to focus on building capacity, rather than getting sucked into the funding vortex. It takes financial resources to sustain the accounting and reporting requirements associated with being a nonprofit. Here again, the usual assumptions about effective science and conservation typically ignore the most important decisions an individual or organization makes. Such weighty decisions as tax structure rarely receive the thoughtful and strategic consideration such profound decisions require.

The Conservation Trap

Conventional conservation has a number of pitfalls or pathologies that limit its effectiveness. These constraints include (1) a lack of vision of how an organization's mission fits in with the broader conservation movement; (2) a narrowly defined mission to save particular wildlife species or pieces of the landscape, rather than working to conserve an integrated whole; (3) turf battles (an unwillingness to work with other organizations with complementary goals); (4) a perceived lack of funding potential for projects and operating expenses translated into narrow, ineffective scopes of work; and (5) various versions of "founder's syndrome," a lack of organizational flexibility because decisions are heavily influenced by the original founder, or the current leadership team persists in imitating the founder's decision-making style. It's a bit of a chicken and egg question to determine whether these pathologies are the outcome of conventional conservation approaches or are precipitated by current pathologies in the funding structure. Likely as not, it is a bit of both. Institutional designs typically avoid considering these issues, assuming they will not occur, and yet they are part and parcel of the process. Considering and anticipating them is foundational to sustaining effective practice; mitigating these intrinsic challenges must be built into program design from the inception.

So how does one organize an approach to avoid these pitfalls? The collective impact paradigm, discussed in chapter 3, provides a structure (fig. 6.7) for how to develop effective science and policy in large, complex systems. The three phases of design, including "initiate action," "organize for impact," and "sustain action and impact," are key to the development of a flexible process capable of evolving to meet changing circumstances. This approach provides a good initial framework for researchers and practitioners alike who are seeking to develop large-scale conservation and science.

As in all the examples that ground this book, stewardship and sustainability are facilitated or destroyed through alignment of economic

Phases of Collective Impact			
Components for Success	**PHASE I** Initiate Action	**PHASE II** Organize for Impact	**PHASE III** Sustain Action and Impact
Governance and Infrastructure	Identify champions and form cross-sector group	Create infrastructure (backbone and processes)	Facilitate and refine
Strategic Planning	Map the landscape and use data to make case	Create common agenda (goals and strategy)	Support implementation (alignment to goals and strategies)
Community Involvement	Facilitate community outreach	Engage community and build public will	Continue engagement and conduct advocacy
Evaluation and Improvement	Analyze baseline data to identify key issues and gaps	Establish shared metrics (indicators, measurement, and approach)	Collect, track, and report progress (process to learn and improve)

Figure 6.7. The phases of collective impact spelled out by Hanleybrown et al. (2012) provide a good road map for practitioners by recognizing relevant action in all three stages, which is key to organizing and sustaining collaboratives of all sizes.

incentives with ecological realities nested in a cultural context (fig. 6.8).[51] What ultimately leads to success or failure is not only the internal program design, but also the external constraints imposed by funders. Do funding strategies inadvertently create perverse incentives or unwinnable frameworks, or set in place the preconditions for sustainability and long-term success? In summarizing the core principles of outcome design, several key points emerge:

- Successful design must be grounded in a clear statement of principles. This includes broad, overarching value statements, as well as more detailed principles of organization, ethics, or practice.

- Guiding principles should include provisions for periodic external review and a clear and transparent process of priority setting, as well as a process for addressing deficiencies if they arise.

- The design must be long term and inclusive, working from the need to sustain programs for multiple decades.

- Sustainability depends, in large measure, upon finance. Strategies need to build local capacity and, to the extent possible, include a plan for reducing donor dependence after the startup period of the program.

- Where long-term donor support is essential, learning feedback loops must include donors, because they often do not learn and evolve alongside the organizations they are supporting.

- Be wary of opportunity costs and indirect effects. Less is more in developing processes and programs that generate resilience rather than debt or other barriers to institutional flexibility. In the end, financial constraints are another scaling issue needing alignment between policy and process.

Concluding Remarks

A science of open spaces addresses the conservation paradox in which small systems are relatively easy to manage but too small to be sustainable, while large systems are ecologically sustainable, but comparatively

Figure 6.8. *Large-scale science and policy is a bit like Russian dolls where researchers typically focus on the smaller internal elements, such as experimental design, while missing the larger social context within which science and policy are embedded. Practitioners typically focus on the larger social context, while missing the integral science and policy elements that provide the critical knowledge feedbacks essential for meaningful conservation. A science of open spaces in general, and resilience design in particular, seek to break down the artificial boundaries between science and policy and show that success for both is completely entwined. (Photo courtesy of Shutterfly.)*

difficult to manage. The approach outlined in the preceding pages demonstrates that effective science and policy are akin to a nested set of perspectives, like a series of Russian dolls, each doll integrated into, and composed of the other, with effective science requiring the development of the preconditions for sustainable action as outlined in resilience design.

This chapter proposes that any plan to sustain large-scale systems must be designed with resilience foremost in mind. As with all complex adaptive systems, a science of open spaces is based on a few ground rules or guiding principles. What grows out of these guidelines is emergent, dynamic, and multifaceted. The aim is to build self-supporting systems in large landscapes using self-interest as a driver for, rather than a barrier to, progress. This is done by attaining and sustaining the organization's vision through approaches that represent ecological and economic win-wins. Without such alignment, durability over the long term is all but impossible.

The science of open spaces addresses the pathologies that lead to loss and implements design principles that encourage resilience and renewal. There are no shortcuts; successful conservation is a multidecade process of seeking solutions, avoiding traps, embracing experimentation, and embodying the lessons from successes, while also recognizing and learning from failure. The preceding pages have provided one framework for considering the complex relationship between humanity and the planet we inhabit. May it help to provide a path forward in more effectively addressing change in one of Earth's most rapidly declining assets—its open spaces.

Notes

(Full citations can be found in the Literature Cited section.)

Preface

1. Lee, 1993.
2. Gell-Mann, 1994; Holland, 1995; Daniels and Walker, 2001; Levin, 1999; Rosenzweig, 2003; Berkes and Folke, 1998; Walker and Salt, 2006, 2012.

Chapter 1

1. See David Western's autobiography, *In the Dust of Kilimanjaro*, 1997, and Western, 2000.
2. Ibid. See also Muchiru et al., 2008; Mwangi and Ostrom, 2009.
3. Funtowicz and Ravetz, 1993; Ravetz, 2002, p. 3.
4. Gell-Mann, *The Quark and the Jaguar*, 1994.
5. See work by Brown et al., 1997, and Curtin et al., 1999, which explore the complex interactions between climate patterns and herbivores.
6. Note that I use the term *fishermen*, rather that the current politically correct phrases *fisher* or *fisher folk* because the fishermen (and women) I know strongly prefer it. A fisher is also a kind of carnivorous mammal, and fisher folk is extremely patronizing.
7. For a review of these issues from Australia, East Africa, and North America, see McAllister et al., 2006; Curtin et al., 2002; Curtin and Western, 2008; Western et al., 2009.
8. Curtin and Western, 2008.
9. See the reviews by Curtin et al., 2002, and Mwangi and Ostrom, 2009, of commons systems in large rangeland ecosystems.
10. See Curtin et al., 2002; Western et al., 2009.
11. Spatial and temporal variability in precipitation may be the single most important explanatory factor influencing landscape vegetation patterns in arid and semi-arid ecosystems. See Muldavin et al., 2008; Augustine, 2010; Sala et al., 2012. At the scale of landscapes, these relationships are complex and can be influenced by numerous other factors, such as interactions with grazers.
12. Curtin et al., 2002.
13. Even in the borderlands, where the establishment of grass banks allowed ranchers to access grass during drought or to restore their own home ranches (Curtin, 2005), this approach has been of limited success because it has typically been short lived, rather than a fundamental transformation in land tenure, and there are costs to moving cattle around, unless it is carefully coordinated and becomes part of long-term grazing strategies. Though unlikely in this generation, for long-term survival, ranchers of the Malpai and elsewhere will probably have to adopt more communal approaches to herding and marketing that will reduce labor costs and increase access to forage across a wider area, while creating better regional coordination of beef production that will facilitate niche marketing of grass-fed beef and higher and more consistent prices per pound.

But this would require a considerable revision in federal and state land regulations to allow multiple owners to graze allotments, a tearing down of fences, and a considerable rethinking on the part of the ranchers of their concept of private property. There is a growing grass bank movement in the West, and this approach is being more widely adopted in Montana and elsewhere. This approach marks a first step toward significant revisions to land tenure that reflect more of the African communal grazing approach.

14. From a social perspective, collaborative place-based conservation is relatively new to North America, though this approach is common in the developing world and is especially well developed in East Africa. Therefore, the process of developing collaborative approaches to resource stewardship is another important lesson to be learned from the African experience.

15. See Muchiru et al., 2008; Western et al., 2009.

16. However, this complex dynamic has in recent decades been threatened by a static view of conservation and land tenure that looks at parks and natural areas as existing without people—the outcome illustrated in the opening pages of this book, where areas without human engagement are often simplified and less resilient.

17. Western, 1997, 2000; Muchiru et al., 2008.

18. D. Western, pers. comm.; D. Sonkoi, pers. comm.

19. See George Hilliard's 1996 history of the Gray Ranch, which provides a valuable window into how land tenure and history intersect on the Diamond A Ranch. Also see Remley's 2000 history of the Bell Ranch, which has one of the few discussions of Spanish and Northern Europe land tenure styles that had important implications for how landscapes were sustained. The Spanish approach that evolved in arid and semi-arid zones was predicated on adapting to variation and was therefore more resilient. The Northern European approach was more intensive and efficient, but also much more prone to collapse. This difference in land tenure is a critical, but little discussed, facet of the history of environmental change in the Southwest.

20. Hilliard, 1996.

21. Remley, 2000; Curtin et al., 2002. See Hess and Holechek's 1995 review of western grazing policy for an additional overview.

22. See Canby, 1981; Hatfield et al., 2013.

23. Remley, 2000; Curtin et al., 2002.

24. Curtin et al., 2002.

25. Hastings and Turner, 1965; Swetnam and Betancourt, 1998.

26. Hess and Holechek, 1995; Curtin et al., 2002.

27. Child and Lyman, 2005.

28. The world's animal population has halved in forty years as humans put unsustainable demands on Earth, the World Wide Fund for Nature's Living Planet Index of 2014 warned (http://wwf.panda.org/about_our_earth/all_publications/living_planet _report/). The dramatic decline in animal species could cost the world billions in economic losses. Humans need 1.5 Earths to sustain their current demands, the report states. The index, which draws on research around WWF's database of 3,000 animal species, is released every two years. The index showed a 52 percent decline in wildlife between 1970 and 2010, far more than earlier estimates of 30 percent. It is due to people killing too many animals for food and destroying their habitats.

29. In the U.S.-Mexico borderlands, ranchers hauling water to stock ponds during drought to preserve endangered leopard frogs were threatened by the prospect that the habitat they created might lead to federal action against them in the event that the popula-

tions inadvertently went extinct. The issue was mitigated by Safe Harbor provisions of the Endangered Species Act, which allowed the development of regional multispecies conservation plans recognizing that if the ranchers were no longer able to haul water, they were not legally required to maintain the population.

30. See Amboseli Conservation Program, http://www.amboseliconservation.org/the -amboseli-ecosystem.html, but information from 2009 is no longer available on the web.

31. See Russell, 2009, and Western et al., 2009.

32. See http://www.soralo.org/about-soralo/.

33. D. Western, per comm., these group ranches are not livestock ranches in the conventional sense, but communal land holdings established in Kenya beginning in the 1960s.

34. http://www.soralo.org/about-soralo/; D. Western, pers. comm.

35. See Corson, 2004, on the ecology of lobster.

36. From a design standpoint, the dynamics of ranching and fishing have far more in common than they do with farming or forestry, which involve fixed assets and completely different approaches to resource use.

37. Wilson, 2002, 2006; Steneck and Wilson, 2010; Curtin, 2010.

38. See the work on local fisheries management and local populations, including Graham et al., 2002, and Ames, 2004. Fisherman Ted Ames was named a MacArthur Fellow for work identifying the local nature of cod populations based on historical fishermen's knowledge, though Graham and colleagues' monograph in many respects provides a more detailed overview of the dynamics of the fish populations.

39. D. Western, pers. comm.

40. Ostrom, 1990.

41. See Corson, 2004, or refer to chapter 3, where these issues are addressed in more detail.

42. See Wilson et al., 2007, for a discussion of the strategic choices made by lobstermen. This example will be discussed in more detail in later chapters.

Chapter 2

1. This phrase was originally coined by borderlands rancher Bill McDonald. It has gone on to influence the work of organizations such as the Quivira Coalition, which seeks diverse partnerships across the West.

2. Brown and Kodric-Brown, 1996; Curtin, 2010.

3. See work by Brown, Valone, and Curtin, 1997, and Swetnam and Betancourt, 1998, that documented vegetation response to rainfall timing, which was thought to be linked to climate change. Balanced between desert and grassland ecotones, southwestern ecosystems are considered to be especially significant bellwethers of environmental change.

4. As researchers working with the Malpai group, our research program flowed directly from research showing that current levels of desertification were climatically driven. Ranches that had not been grazed for decades showed accelerating rates of change, but they also showed two to three times the change of grazed landscapes, indicating that large grazers such as cattle are key for mitigating climatic effects (see Curtin and Brown, 2001). The research programs in the borderlands expanded on this preliminary evidence to examine how communities through active grazing and fire could adaptively manage to address climate change impacts.

5. See Maestas et al., 2002, paper on the impacts of exurban development on wildlife. Though focused on Colorado, the results are equally relevant to the borderlands.

6. See Curtin et al., 2002, for a review of range science in the borderlands and its interface with larger and more dynamic approaches to conservation and scholarship.

7. This was, perhaps, a pragmatic decision of the MBG leadership, who recognized that range scientists, with their long association with ranchers, had less credibility with the mainstream environmental community.

8. See Wolf, 2001; Curtin, 2002a, 2005; Sayre, 2005.

9. Bill McDonald and much of the group very much resisted the title of "model" placed on them by agencies, conservation organizations, funders, and others. They saw what they did as necessary and effective in their landscape, but not necessarily relevant to others. But as the visits and press increased, the group came to accept and assist in informing other groups, including ranching workshops for other parts of the country that spawned now successful organizations across the West, as well as exchanges with East African and Mongolian herders, and outreach to groups in the Middle East and even entirely different systems such as fisheries, as chronicled in the opening sections of this book.

10. The Portal Project has become one of the longest running and most widely cited ecological experiments on the continent. It was begun by James H. Brown and colleagues in 1977. By the time I worked on the project in the mid-1990s, the nearly twenty years of data were profoundly influencing the way ecologists approached complex interactions between organisms. See Brown et al., 1986, for an early overview of the project and its experimental design.

11. The phrase "wicked problem" was originally used in social planning and stems primarily from the work of Rittel and Webber (1973). It describes problems without clear solutions or definable end points, an increasingly common reality in conservation and resource management. We will return to the concept and its implications in later chapters, for the reality of this intrinsic complexity and how to address it is a cornerstone of this book.

12. For example, Milchunas's 2006 review of grazing studies in the southwestern United States documented that out of hundreds of peer-review studies, virtually none are conducted at the scale where grazing actually occurs on the landscape. In these scale-dependent systems (and grazing impacts in particular are very much an outcome of how much room the cattle have to roam), viewing ecological processes at a level consistent with management completely changes research implications and has profound implications for land tenure and developing sustainable land use.

13. Milchunas and Lauenroth's 1993 monograph, comparing grazing studies through a global meta-analysis, remains one of the definitive analyses of grazing impacts. Over the course of more than fifty studies, there was little documentation of negative impacts, with the exception being times of drought. Though data existed to support the premise that grazing was more damaging during drought, less was known about the thresholds of change in grazing impact in relation to climate, which in many management situations is the key question. Many researchers and practitioners still hotly debated whether grazing in arid and semi-arid climates was compatible with conservation at all. The Nature Conservancy and researchers working with the MBG took a good deal of heat for collaborating with ranchers.

14. The British mathetician George Box famously stated, "All models are wrong, but some are useful." Such is the case with the Malpai model of mid-elevation interactions (figure 2.4), where the simplifying assumptions allow one to see more clearly the fundamental components of the system.

15. The monitoring was done for a range of reasons, from assessing range health on the vast Gray Ranch as part of an easement with The Nature Conservancy, to documenting fire effects on plant and animal species of special concern, including agave and bats, montane rattlesnakes, and jaguar.

16. The term peer-review-quality science appears in frequent meeting notes and e-mail correspondence during the group's formative period. The term was used at the urging of John Cook and science advisors such as Jim Brown and Ray Turner and was adopted as the group's standard.

17. In the 2005 White House Conference on Collaborative Conservation, of the more than 400 groups represented, the MBG was the only one to experimentally test their underlying assumptions, and one of a very few to conduct coordinated research of any kind. See biologist Karl Hess's summary of collaborating groups in the United States (http://govinfo.library.unt.edu/whccc/).

18. See Wolf, 2001.

19. See Funtowicz and Ravetz's (1993) classic discussion of reconceiving science, as discussed in the introductory chapter.

20. Recent articles in the journal *Science* have shown that close to 80 percent of the results of laboratory studies are not reproducible. Yet reproducibility is supposed to be the gold standard of science. With enough design and statistics, even contradictory hypotheses can both be proven true. In the words of Mark Twain (paraphrasing nineteenth-century British Prime Minister Benjamin Disraeli): "There are lies, damned lies, and statistics."

21. Ecologist David Schindler and colleagues, and comparable work by Steve Carpenter and associates, independently demonstrated the shortcomings of microcosm studies and how small studies cannot be scaled up to larger environments. However, the approach has remained commonplace in ecology.

22. A senior researcher at the U.S. Department of Agriculture's Jornada Experimental Range—the premier range ecology research unit in the Southwest—once told me that I would never get meaningful results by working at such large scales, there was simply too much "noise." My response was that if one cannot detect patterns through the noise then it's likely that one's results may be statistical anomalies or simply artifacts of experimental design. Though harder to publish and certainly less productive in terms of paper generation than conventional approaches favoring microcosms over macrocosms, the whole idea was to see what emerged from taking a "crude look at the whole."

23. The Gray Ranch was recently renamed the Diamond A Ranch and expanded by half again in size. The Gray Ranch was The Nature Conservancy's first metaproject in the West begun in the 1980s. TNC still holds the easement on the ranch that was purchased by the Animas Foundation in the early 1990s.

24. In a contrast mimicking the debate over "simple" versus "systems" theorists of the 1960s, the bias toward precise, rather than accurate, answers remains. Ironically, some of the harshest critics were from a local land grant university whose mission was to be relevant, illustrating how deep these biases extend. Senge (1990) made the additional point of the significance of lag effects in these systems, another factor that is missed in simple and short-term studies.

25. In addition to Gottfried and colleagues' 1999 state of the knowledge review, Curtin et al., 2002, provided an overview of the conservation value of ranching across the Intermountain West. Also see Curtin, 2008; a monograph of the initial results of a decade of experimental studies in the borderlands.

26. See National Research Council, 1994; Brown and McDonald, 1995; Curtin, 2002b; and Curtin, 2008. But also see Jones, 2000, for a different perspective.

27. Curtin, 2008, documented the role of pronghorn in influencing vegetation composition and the disproportionately large role even relatively rare native grazers can have in sustaining vegetation diversity.

28. In addition to research from McKinney Flats, much of the evidence comes from a Mexican National University team led by Rurik List, who documented widespread bison populations prior to settlement, indicating that large grazers were a significant contributor to desert grasslands (List et al., 2007).

29. See work by Cable, 1965, and related studies. For a good review of fire in desert grasslands, see McClaran and Van Devender's 1995 book on desert grasslands from University of Arizona Press, especially Guy McPherson's excellent chapter on fire effects. The work is interesting in that it comes just prior to the paradigm shift to considering fire as healthy in desert grasslands, but provides an extensive review of the literature and a good summary of the pre–paradigm shift perspective.

30. See Curtin, 2008, and associated papers. In our studies even intensive grazing had little or no impact on grass, invertebrates, and small mammals.

31. Ibid.

32. Through what are euphemistically called "patty counts," the amount of cattle manure can be quantified as a rough index of relative use of a given area.

33. See Curtin, 2008. Especially striking were the direct and indirect impacts of pronghorn (*Antilocapra americana*) and their interactions with fire. Even in small numbers, pronghorn were able to flip the climate-fire interactions.

34. See Truett et al., 2001. Truett, a biologist who spearheaded prairie dog restorations for the Turner ranches, noted this pattern himself and uncovered nineteenth-century accounts of prairie dog populations crashing with the loss of bison, or increasing with the introduction of cattle.

35. Again see Curtin, 2008, for a review of the process in action. These are just the initial results of continuing data analysis.

36. For example, early on in my work for Cascabel, I found that the mammal sampling I was conducting, intended to be compatible with that of McKinney Flats by studying the effects of rodents on vegetation, wasn't working; the rodent population was too small to attain effective samples on the higher elevation site. I suggested we discontinue the mammal research and instead use the funds to analyze existing data and create more synergy with other parts of the borderlands research program. Mammal sampling was accordingly stopped. However, the newly liberated funds were simply redirected into intensive vegetation sampling undertaken by colleagues of the principal investigator (PI) (a separate program using the same property) that was redundant to our other ongoing experiments. One morning, my colleagues and I arrived at our site to find the entire research area festooned with the PI's fresh ribbons and markings, making it nearly impossible for us to locate our own research plots. Instead of sufficiently supporting a single line of vegetation sampling, the Forest Service decided to add a second sampling protocol right on top of the existing one. This meant that both Cascabel and the PI never had the resources to fully analyze the data, nor could we make the best use of the resources we had. Though perhaps a reflection of the PI's approach, the same general pathologies have been seen in federal programs across the borderlands, where there was frequently little continuity in the science.

37. The expression "doing the wrong things, righter" is a reference to the wonderful piece on failed governance approaches by Australians Ison and Collins, 2008.

38. Though federal taxpayer funding is supposed to be transparent and open to public review, the reality of the way the books are kept makes it very difficult to determine how funds were actually spent. Rough estimates suggest the yearly cost of Cascabel

must have been at least twice that of McKinney Flats, so Cascabel probably incurred well over a million dollars of expenses in the decade of the project.

39. Curtin, 2002, 2005, 2008, 2010; Curtin and Western, 2008.

40. Keep in mind that although the Cascabel experience is pretty typical, it does not represent all federal science. Many federal research projects are extremely effective at generating cutting-edge research and promoting broader understandings.

41. This perspective is opposed to the narrow focus on species or economic or legal interests still typified by many conservation agencies and organizations at the time and which continues to be a large part of conservation biology.

42. See predictions by CLIMAS (Climate Assessment for the Southwest) based out of the University of Arizona (http://www.climas.arizona.edu/) or the recent fine analysis undertaken by Patrick McCarthy (2006) and collaborators at The Nature Conservancy and related organizations.

43. Note that these actions are in stark contrast to those of managers of many of the West's large ranches, which welcome research. The Turner ranches are a case in point, where researchers are an asset to be cultivated, not a problem to be mitigated. As one Turner ranch manager noted, "It's free information, who would not want it?"

44. A 1998 letter from the Animas Foundation stated the intent to sustain the experimental program on McKinney Flats for "decades," and that the work was the cornerstone of the organization's goals for science-based management.

45. The Gray Ranch was a pioneering effort by The Nature Conservancy in promoting science-based ranching. Although limiting access by outside people and the widespread use of fire have restored much of the range, there seems no question that the project has fallen well short of its potential. The changes on the Diamond A not only compromised the MBG's mission, they raised larger questions about the long-term efficacy of conservation-minded ranching. The establishment of the Animas Foundation to manage the Gray/Diamond A Ranch had once been viewed as a best-case scenario for conservation, and the Animas Foundation as a flagship organization in promoting science-based stewardship and the use of ranches to promote science and conservation.

46. As demonstrated in the pioneering work of Aldo Leopold, trapping predators is rarely effective. The trapping on the Diamond A was especially flawed, because many areas they trapped were immediately adjacent to Mexico, so predators still flowed in from adjacent landscapes. The relationship between pronghorn survivorship and predators was never documented and not supported by most wildlife managers.

47. Ranch records from the time of ownership by Mexico billionaire Pablo Brener showed a sophisticated livestock management approach with extensive rainfall and vegetation monitoring.

48. However, we also realized that this is a ranchers' organization and to grow and sustain the organization required greater rancher engagement. Not all the ranchers understood the significance of the science or shared Bill McDonald's and President Reese Woodling's enthusiasm for it.

49. Though most dictionaries define science as research involving experimental, hypothesis-driven approaches, across the West, the term has been diluted to mean essentially any form of data collection, a process that is vastly different from the peer-review-quality research envisioned at the start of the program.

50. The reasons for this are numerous. After a decade of research we came to realize which procedures yielded the most information, and the reduction in startup costs led

to increased efficiencies in monitoring. As biologists gained experience, their salaries tended to increase, and there was a reluctance to abandon previous protocols, even when they had been shown to be ineffective. Partly this results from the need for continuity in data collection. But much of it was also just a reflection of a reluctance to critically evaluate the effectiveness of existing protocols.

51. A monograph by Gottfried et al., 1999, summarized the early research in the borderlands in support of conservation and management.

52. The findings of a blue-ribbon panel on rattlesnake conservation that addressed issues concerning rare rattlesnakes and fire were taken into consideration for prescribed fire planning processes in 2003. This constituted a good example of how effective this approach could be.

53. The group periodically engages scientists from the USDA Jornada Experimental Range and similar experts. Although the researchers are globally recognized authorities in range management, they are still relatively local and long-term collaborators of the group. This process is different from engaging a series of experts from outside who view the system through entirely fresh eyes. The problem is that so much of the review comes from people who in one way or another have a stake in the process, well meaning as it may be, and this influences the type of insights one receives from the process.

54. A case in point is a paper by one long-time borderlands researcher who called into question the effectiveness of recovery following certain kinds of management actions. The conclusions, though perhaps not distributed in the most politically astute manner, caused a sharp backlash by the ranchers, who were used to having science work consistently in their favor. However, the researcher was only doing his job.

55. As advisors came on they tended to "go native," either focusing on their own interests or saying what they thought the ranchers wanted to hear. This left independent scientists, such as myself in the position of frequent naysayers, when all we were trying to do was live up to our ethical responsibilities and give the group sound advice as we had committed to do at the outset of the project.

56. The savanna managed by the U.S. Forest Service as part of the Cascabel Project, which showed such promise in complementing the lower elevation grassland project of McKinney Flats quickly became a political football that drew energy and resources away from the core Malpai science program. The Cascabel experience poignantly highlights the importance of broad external oversight and how a systematic yearly review of all the programs could have prevented poor management and waste of resources.

57. The very use of the terms "ecosystem management" and "scientific uncertainty" in the title suggests the author did not know that the whole point of ecosystem management is to embrace and learn from uncertainty and promote relevant science. See the classic papers on adaptive management by Holling and colleagues cited later in this volume, and the discussion in chapter 5 (Holling, 1978; Lee, 1993; Gunderson et al., 1995).

Chapter 3

1. Former Maine Marine Commissioner Spencer Apollonio wrote a fascinating book applying hierarchy theory to the marine ecosystems of Maine (2002). In the book are numerous quotes from the time of European contact discussing the immense richness of the system. For example, a 1639 quote stated, "The abundance of sea fish was almost beyond believing, and sure am I should scare [*sic*] have believed it has [*sic*] I not

seen it with my own eyes." However, archeological evidence by Bourque (1995) suggests the system had already been considerably fished-down by Native Americans in the thousands of years prior to Europeans arriving on the shores of the New World.

2. See Colin Woodward's wonderful history of the coast of Maine, 2005.

3. For a review of the ecology of lobser, see Corson, 2004.

4. As noted in chapter 1, I use the term "lobstermen" or "fishermen," rather than "fishers," or "fisher folk," because fishermen (and women) prefer it.

5. See the discussion of gilded traps in Steneck et al., 2011.

6. See Corson, 2004; Robert Steneck, University of Maine, pers. comm.

7. Traps are now required by law to have an escape door that rots open so the lost "ghost" traps do not keep fishing.

8. See Corson, 2004.

9. Acheson's *The Lobster Gangs of Maine* (1988) is the classic study of distribution of resource use through local self-organized governance structures organized around local harbors.

10. See Wilson et al., 2007, discussed in chapter 1, for additional details.

11. Based on conversations with lobster biologist Diane Cowan, director of the Maine-based Lobster Conservancy.

12. A mobile dynamic had existed in the fishery for at least seventy or eighty years as steam and later gas and diesel engines allowed boats and the associated economy greater mobility.

13. The larger boats have mostly moved to the Massachusetts port of Gloucester and other areas closer to the offshore banks and where they have more support infrastructure. In Maine, where there are more than 7,000 registered lobster fishermen, the fleet of draggers is at most a few dozen, down from hundreds a few years ago.

14. See Wilson, 2002, for a wonderful overview of a complex systems-based approach to fisheries.

15. See Berkes et al., 2006, for a review of exploitive approaches to fisheries and other forms of natural resources management.

16. See Wilson, 2002.

17. See Alverson et al., 1996, and Harrington et al., 2005.

18. See Wilson, 2006, for a review of the complex relationship between fisheries and the scale of resource management, with the crucial take-home message that the ecological scale of the system and the scale of management must match, or destruction of resources occurs.

19. See Wilson, 2002.

20. Ibid.

21. Ibid.

22. See Sissenwine, 1984, and Walters, 1998, for a review of the effectiveness of stock assessments and how they are notoriously inaccurate.

23. Quoted from Smith's 1990 article on chaotic ocean systems.

24. See Apollonio, 2002.

25. Some fishermen did challenge the underlying assumptions of the approach by recognizing greater complexity within the system (Smith, 1990), although these arguments were never officially recognized by the scientists or resource managers because they did not fit the paradigm of stock assessment. Even in the 1990s, as environmental organizations successfully challenged National Marine Fisheries Service policies and amendments to the Fishery Conservation and Management Act, these challenges never questioned the underlying validity of the single-species approach.

26. See classic articles by Pauly et al., 1998, Jackson et al., 2001, and Myers and Worm, 2003, on global decline of fisheries.

27. See Bourque's 1995 review of long-term changes in the Gulf of Maine based on archeological evidence, also summarized in part in Jackson et al., 2001.

28. See Kurlansky, 1997, for a wonderful review of the role of cod in Atlantic ecology and culture. Also see O'Leary's classic 1996 book on the rise and fall of Maine fisheries.

29. The purse seine was a long net which, when a school of fish was encountered, could be launched from a dory and tightened around the school. Although the purse seine was deployed after the fish were found and could selectively harvest fish, if they were not the target species, the fish could be released unharmed. The otter trawl was much less discriminating in its capture of fish. It was essentially a large net dragged through the water column. It allowed much larger harvests, for fish were captured even when they were not actively foraging, such as during spawning periods, where there were high densities of fish in nearshore waters.

30. The towing of nets through reefs and other delicate habitats also homogenized bottom habitats. This reduction in habitat diversity is also considered to have reduced the health of marine ecosystems.

31. Ames, 2004.

32. Personal communication with former Maine Marine Commissioner Robin Alden, who relayed the hazards of developing new marine policy in an era of contentious political debate.

33. The Cobscook Bay Resource Center model of local conservation, developed in the Canadian maritime provinces, explicitly links local ecology and economy in a unique place-based approach that seeks to empower fishermen, while also better educating them about sustainable approaches to fisheries. The approach holds great promise for terrestrial systems. In some ways groups like the MBG function very much like a resource center, but the model has yet to be explicitly exported to land. However, the Cobscook Bay Resource Center also increasingly works with local farmers as well as fishermen. See Kearny, 2005, for a review of the Canadian approach to community fisheries.

34. See Kania and Kramer, 2011, and Hanleybrown et al., 2012. Collective impact is in many respects the most recent paradigm in the funding world; one that puts together in a clearly understandable format many of the best practices from collaborative conservation and community development, but doing so in a framework that potentially increases the scale and efficiency of large-scale conservation programs.

Chapter 4

1. Such an assumption is not new, as British astrophysicist Arthur Eddington asserted some eighty years ago:"The second law of thermodynamics—holds, I think, the supreme position among the laws of nature. If someone points out to you that your pet theory of the universe is in disagreement with Maxwell's equations—then so much the worse for Maxwell's equations. If it is found to be contradicted by observation—well, these experimentalists do bungle things sometimes. But if your theory is found to be against the second law of thermodynamics I can give you no hope; there is nothing for it but to collapse in deepest humiliation."

2. See classic work by Schnieder and Kay (1994), which reformulates the second law of thermodynamics, thereby making it much more relevant to biological and social systems. Previously the law applied only to physical systems.

3. The role of energetics in organizing ecological systems has been considered for decades (e.g., Margalef, 1968; Odum, 1983; Brown, 1994).

4. Jorgensen and Fath, 2004.

5. See Choi et al., 2005, for a discussion of devolution in the Scotian shelf ecosystem.

6. Ecological extinction means a species can still be present in the ecosystem, but it exists at such low levels as to no longer serve its historic function in the ecosystem. See a classic paper by Jackson et al., 2001, which considers global changes in fish population.

7. This is important because it explains why adaptive capacity—the ability to adapt to change—is crucial in sustaining communities and ecosystems. Adaptive capacity is a balance between social and ecological processes and requires a diversity of component parts, or requisite variety, as a cornerstone of sustainability. "Requisite variety" is needed in all systems to maintain or maximize the component parts of large systems. It encompasses factors ranging from genetic variation to biodiversity that contribute to a system's ability to respond to change. This is also known as the law of requisite variety or Ashby's law, after early systems and cybernetics thinker William Ross Ashby (1962), who stated, "The larger the variety of actions available to a control system, the larger the variety of perturbations it is able to compensate."

8. There is some debate in biological circles about how ecological systems are kept in control. Those who favor the idea of top-down control hold that the top-level predators keep the numbers of herbivores in check. For example, according to this theory, lemming populations in the Arctic are controlled by the numbers of predators, such as hawks, owls, and arctic foxes, that are around to eat the lemmings. In contrast, advocates of bottom-up control hold that the availability of the lemmings' own food supply controls the lemming population, and that the size of the lemming population dictates the size of the predator population.

9. The classic example is ecologist Aldo Leopold's studies on the Kaibab Plateau in Arizona, where elimination of predators caused a dramatic increase and then collapse of deer populations. See Leopold, 1933.

10. The point is somewhat contentious, with the impact of wolves documented in some areas and not in others. For an article explaining the trophic cascade in more detail, see Ripple and Beschta, 2011. For evidence of lack of impact, see Kauffman and colleagues' 2010 article.

11. For specific details, see Bourque, 1995; Jackson et al., 2001; and Steneck et al., 2004. The pattern of change is worldwide, as illustrated by the work of Pauly et al., 1998; Jackson et al., 2001; and Scheffer et al., 2005.

12. See Steneck et al., 2004, for more details.

13. This interpretation has been hotly debated. Although Steneck and colleagues argued that materials from sea urchins and other nonbones are not present, suggesting that lobster were not present, others argue that lobsters do in fact rot more easily and that this explains their absence in the archaeological record, rather than scarcity due to predators.

14. See Bourque, 1995, and Jackson et al., 2001. The best data come from the Turner Farm site on North Haven Island, Maine, which through a more than 4,000-year chronology illustrates a shift in species composition and size, indicating that a classic fishing-down of the ecosystem happened well before Europeans arrived on the scene.

15. See note 1, chapter 2.

16. The same pattern holds true for terrestrial systems. An eighteenth-century analog was the harvest of beaver in the New World to satisfy global markets. The loss of beaver as a crucial engineer species transformed the ecology of the region and the economies of the globe. See Innis's classic work (1977) on the Canadian fur trade.

17. The term "novel ecosystems" used by Australian ecologist Richard Hobbs and colleagues (2006) refers to fundamental changes in ecological structure that have occurred in response to human action. Although ecological systems are always evolving new structures in response to change, the rate of change and the development of these systems increased dramatically in recent decades.

18. See Myers and Worm's meta-analysis of nine continental-shelf ecosystems (2003) and other reviews of nearshore ecosystems, including Frank et al. (2005, 2007), and Scheffer et al. (2005).

19. Though lobster, as with shrimp and snow crab in Canada, are of far higher economic value then the fisheries they replaced, this creates the gilded trap discussed in chapter 3, wherein higher economic returns lead to greater capitalization of the fleet. This system is also intrinsically less stable due to the potential for disease or stochastic events, such as shifts in currents changing recruitment success, that potentially led to complete economic collapse of much of the coastal economy.

20. Ted Ames, pers. comm.

21. The instabilities are not just ecological, but also economic. In 2008 the collapse of the Icelandic economy led to a loss of funding for Canadian canneries that use excess capacity of U.S. lobster. Those lobsters that cannot be sold locally, or that are not of a quality to be shipped live, are sold to Canada, to be primarily resold back in the United States. Much of the "Maine" lobster found in restaurants is actually from the Canadian canneries. The loss of this market collapsed the wholesale price to near $2.00 per pound.

22. Fishermen have noted dramatic changes on the coast of Maine, where in less than the space of a single human life the diversity of nearshore waters has been profoundly reduced.

23. Internal and external factors were also associated with ecosystem transition. Physical environmental changes may also have contributed to the pattern, with shifts in deep-water temperatures and ocean acidification potentially contributing to the diminished energy flux in the benthic fish community, as revealed by reduced physiological condition and reproductive output of midsize predatory fish (Frank et al., 2005). Regardless of the precise cause, there continues to be a question of whether recovery and system reassembly can come from the recovery of cod, or whether more fundamental adjustments in the ecosystem are needed to fuel the processes.

24. This same pattern is exhibited in terrestrial systems, such as in the U.S. Southwest or other arid and semi-arid ecosystems, where overgrazing also leads to a punctuated decline in resources.

25. See Brown et al., 1997, Swetnam and Betancourt, 1998, Curtin and Brown, 2001, and Curtin, 2008, for more details on the outcomes of environmental change in the Southwest.

26. See Brown et al., 1997.

27. Also known as river herring, the name alewife is said to stem from descriptions of a large rotund woman or tavern-keeper's wife. The fish's belly and shape are stouter than similar forage fish.

28. See Curtin and Hammitt, 2012.

29. Note that fishing impacts are all relative. When seines were first implemented, they were thought to destroy the fishery, but now they seem relatively benign. Yet in reality even these techniques may be too much, and fisheries may need to return to older and even simpler methods, much as the lobster fishery has sustained.

30. See Manchester, 1992, and Worster, 1994.

31. The classic example is Mandelbrot, 1967, in which he shows that the length of area is self-similar and that length is largely a reflection of the scale at which you view the question.
32. Baranger, 2011.
33. For a review of the foundations of cybernetics and systems thinking, see work by Boulding, 1950; Ashby, 1962; and von Bertalamffy, 1968.
34. See Hilborn, 2004, who discussed how the butterfly effect, in which insects influence weather by manipulating initial conditions, originally dates to papers from the 1800s. However, this literature, written in French, was unlikely to have been known when the concept was coined in the 1970s.
35. See Holland's 1995 review of complexity.
36. See discussions of complexity in social and ecological systems by Taylor, 2005; Wessels, 2006; and Norberg and Cumming, 2008.
37. See discussion of work by Wilson et al., 2007, reviewed in the previous chapter and in chapter 1.
38. See Tobey, 1981, which reviews the history of Clements and the Nebraska School.
39. Ibid.
40. Clements's student Paul Sears advised the Roosevelt administration and for decades was influential in North American conservation. His work set the stage for many of the current integrative approaches to sustaining the environment. Sears went on to be especially influential through his landmark book, *Deserts on the March,* 1935, which is still considered an authoritative text on desertification.
41. Clements was a Lamarckian who did not recognize evolutionary processes in his perceptions of the environment and ecology. Lamarck is best known for his theory of inheritance of acquired characteristics, first presented in 1801: If an organism changes during life to adapt to its environment, those changes are passed on to its offspring. Lamarck said that change is made by what the organisms want or need. (Darwin's first book dealing with natural selection was published in 1859.)
42. The author got to know many members of the MacArthur family firsthand as a college student, and was inspired by their range of interests and expertise that no doubt contributed to MacArthur's innovative approach. Also see Wilson and Hutchinson, 1989. Though still favored in many ecological sciences, single-species or population-based approaches are now being challenged by broader and more dynamic perspectives (T. H. F. Allen, pers. comm., 1994; J. H. Brown, pers. comm., 1995; C. S. Holling, pers. comm., 1998). Ecologists are recognizing that ecological systems are not merely the sum of their parts, but rather have emergent synergies that must be examined as a whole (Holling, 1986; Holland, 1995).
43. See Brown, 1999.
44. MacArthur died prematurely, at age 42, of renal cancer, in 1972. Many have speculated how his influence on ecology may have changed if his career had lasted another thirty years.
45. The word "ecosystem" can be traced to a 1935 paper by the British ecologist Arthur Tansley. The first full paper dealing with ecosystems was a 1942 study of bogs by Raymond Lindeman.
46. Leopold was unique in linking science with ethics. He realized that just below the surface there are always value judgments and ethics and that many of the scientific battles over ideas are often using data as a proxy for values. Much of science does not recognize that the two are not reconcilable. Granted Leopold's *A Sand County Almanac* (1949) was a series of popular essays and not a research article, but the exponential

growth in references to it through the decades highlights that it fills a niche largely unaddressed in later writings. The quote is from the final essay, "Land Ethic."

47. See Frank Golley's 1993 review of the history of ecosystem ecology.

48. The Odum brothers wrote ecology's first textbook, *Fundamentals of Ecology*, published in 1953. Its emphasis on physical systems was extremely influential in broadening the perspective of the field.

49. See Wilson and Hutchinson's 1989 tribute to Robert MacArthur.

50. See Brown, 1999; Wilson and Hutchinson, 1989.

51. Ibid.

52. The concept extends back more than eigthy years to Cain's 1938 article, through classic work by Preston, 1962, and MacArthur and Wilson, 1967, and formed a cornerstone of early writing in conservation biology (e.g., Soulé and Wilcox, 1980) and is still influential to this day.

53. See Michael Rosenzweig's 2003 book, which spells out ways of sustaining the environment while also meeting the needs of humanity. Also see Rosenzweig, 1995, for an overview of ecological issues of diversity.

54. See David Quammen, 1996, *The Song of the Dodo,* which discussed the roots of biogeography.

55. See Allen and Starr, 1982, and O'Neill et al.,1986, for reviews of the ecological foundations of hierarchy theory.

56. See Herbert Simon's classic article "Architecture of Complexity," 1962, which was among the first to propose means of organizing complex systems.

57. O'Neill et al., 1986.

58. See Allen and Starr, 1982.

59. For more discussion of these perspectives, see Allen and Hoekstra's landmark book, *Toward a Unified Ecology,* 1992, which uses a review of the discipline as a foil for exploring distinctions between scale and type and the richness of a worldview that explores complex systems at a range of levels.

60. See Curtin, 2010, for a review of scale impacts on marine and terrestrial systems.

61. Again see papers by Wilson, 2006, Steneck and Wilson, 2010, and Curtin, 2010, reviewed in previous chapters.

62. Lee, 1993.

63. This ties back to foundational principles of cybernetics. Known as Ashby's law (1962), the more options or "requisite variety," the more opportunities there are for innovative problem solving.

64. Core to Bateson's concept of memory is the cybernetics idea of nonlinear feedback loops. In cybernetics a circular connection between causal components eventually feeds back to the point it started, to begin again. This concept is significant in how it informs different approaches to learning, discussed later in the chapter.

65. Capra, 1996, synthesized the systems theory literature and, in particular, Maturana and Varela's contribution, by setting out three criteria for a living system: the pattern of organization, the structure, and the life process. These include three points: (1) Pattern of organization is the configuration of relationships that determines the system's essential characteristics (autopoiesis as defined by Maturana and Varela, 1987). (2) Structure is the physical embodiment of the system's pattern of organization (dissipative structures as defined by Prigogine and Stengers, 1984). (3) Life process is the activity involved in the continual embodiment of the system's pattern of organization (cognition as defined by Gregory Bateson, 1979).

66. See reviews of cognition by Maturana and Varela, 1987, and Beratan, 2007.

67. See Cahill and McGaugh, 1998.
68. See Beratan's 2007 review of cognition as it relates to natural resource management.
69. See Hutchins, 1995.
70. Maturana and Varela, 1987.
71. Proulx, 2008.
72. Hutchins, 1995.
73. See the classic work of March and Simon, 1958.
74. When fishermen are given the opportunity to see the ocean floor through rides in submarines, they are often struck by how much it looks like they thought it would. It often seems they know the contours of the landscape they work from above as well as ranchers or farmers who transverse a landscape and see it with their own eyes.
75. Lee, 1993.
76. Bargh and Chartrand, 1999.
77. Simon, 1947.
78. See Tainter's classic work (1990) on the collapse of complex societies. A classic example is the decline of the Roman Empire, where cultural norms and concentrated power eventually led to brittleness and vulnerability, even when they were counterproductive to preserving overall well-being.
79. Coronado National Forest supervisor, Douglas District, 2003.
80. Building on Simon (1947), Cyert and March (1963) cited four concepts that relate to the function and collective learning of institutions such as the Forest Service: (1) There is local rationality with goals independent of constraints. (2) There is uncertainty avoidance. (3) Searches for solutions are problematic if they are oversimplified or biased. (4) Organizational learning involves adaptation to goals and attention to rules, while being in search of effective new rules.
81. Curtin, 2014, discussed the process of resilience design (addressed in the next two chapters of this book), in which consideration of sustainable process is built into the governance of the system from the outset, just as the Malpai Borderlands Group built in science as a framework for decision making.
82. Argyris and Schön, 1978.
83. Ibid.
84. Armitage, 2008.
85. See Holling and Meffe's 1996 article on command and control approaches to governance, which are rigid, directed approaches. The cost and benefits of different governance structures will be revisited in the next two chapters. Also see Kai Lee's classic work, *Compass and Gyroscope,* 1993 (discussed in the next chapter).
86. See Flood and Romm, 1996.
87. Morgan, 1988; Wang and Ahmed, 2001; Leeuwis and Pyburn, 2002; Keen et al., 2005.
88. *Umwelt,* from German meaning "environment" or "surrounding world."
89. Some authors even propose a fourth loop of learning that is even more reflective and philosophical by evaluating the foundation justifications and logic (Loverde, 2005), though from the point of view of conservation design, the essential point remains the same.
90. Curtin and Western, 2008.
91. See Easterby-Smith et al., 1999, on organizational learning.
92. Ostrom, 1990, 2007; and Holling and Meffe, 1996.
93. See foundational papers by Holling, 1973, 1986. The concept of resilience is addressed in detail in the next chapter.
94. As stated earlier, the perception is also akin to geneticist Sewall Wright's concept of

adaptive landscapes (1932), in which fitness is maximized through attaining far-from-equilibrium adaptive peaks.

95. An understanding of cognition and institutional design strongly suggests that a primarily collaborative, place-based approach provides the institutional resilience and durability sufficient to address change. This does not mean that bottom-up approaches are necessarily better, but that a dichotomy between top-down and bottom-up essentially asks the wrong question, for successful policy will have elements of both. Grounding in place and developing the processes that allow "learning to learn" are key ingredients of sustainable science and policy.

96. Westley et al., 2006; Westley, 2008.

97. See review of innovation by Mumford, 2002.

98. See Westley, 2008.

99. Morgan, 1988.

100. Westley, 1995, in Gunderson et al., 1995, *Barriers and Bridges.*

101. Westley, 2008.

Chapter 5

1. Theorists Jantasch, 1980, and Prigogine, 1996, both recognized how much the creative elements of their work were embedded in a larger societal context.

2. See Masten et al., 1990, for a review of the psychological disciplines approach to resilience.

3. Recall the papers by Jim Wilson, 2006, and Steneck and Wilson, 2010, that have eloquent discussion of the scale mismatch in fisheries management.

4. The term *socioecological system* has come into vogue in recent years, but as explained in earlier sections, the approach tends to involve parallel play by different disciplines, rather than an integrated approach, as is really needed to solve environmental problems. The work of Daniel McCarthy, 2006, on linked socioecological systems is some of the most effective discourse on the topic.

5. The currency of science is paper production—the maximization of what some scientists cynically call MPUs (minimum publishable units). Balanced with this is the need to have work widely cited and published in more prestigious journals. This rule set is extremely effective at promoting academic careers, but less effective at promoting relevant knowledge or long and thoughtful approaches to tackling major problems. Consider that Charles Darwin sat on the ideas of evolution for decades, building the evidence. Such a long, protracted, and thoughtful approach would be impossible today. Charles Darwin arguably would never have gotten tenure. See *The Economist* (http://www.economist.com/news/leaders/21588069-scientific-research-has-changed-world-now-it-needs-change-itself-how-science-goes-wrong) and *Nature* (Nature, 2011).

6. This tension over control and certainty versus complexity and flexibility has marked tensions in science versus religion extending back to René Decartes and the tension between faith and reason. Science and reason focused on control and predictability, and emergent phenomena were left to the realm of faith. Increasingly science does not see these two approaches as incompatible, just as nonequilibrium approaches, too, are not new, but have also gained sway in an increasingly complex and nuanced worldview. See Kaufman, 2008.

7. See Holling, 1959, as an example of his early, yet classic, work in populations interactions. Holling was remarkable in producing seminal papers in both population and systems ecology.

8. In his cadre were researchers from a diverse range of backgrounds. Holling's graduate student William C. Clark joined the Ecological Policy Group in 1971, where he worked until 1980. Clark pioneered work in ecological policy design, became a professor at Harvard, was named a prestigious MacArthur Fellow, and was a founder of sustainability studies. Other members of the UBC group included fisheries biologists Ray Hilborn and Carl Walters, mathematician Don Ludwig, and systems analyst Dixon Jones, all of whom made major contributions across a range of disciplines. See Clark et al., 1975, 1979; Holling, 2006; Walters and Hilborn, 1975, 1978; and Gunderson et al., 2010. For a more condensed review of the foundations of resilience, see Curtin and Parker, 2014.

9. In a 1998 conversation, Holling remarked, "it all goes back to Leopold," acknowledging that his work was built upon Leopold's foundation.

10. For the original foundations of theory, see MacArthur, 1955, discussed in this chapter and in chapter 4. For an example of wholesale ecosystem collapse in response to climate change, see Terry Hughes, 1994.

11. Australians Walker and Salt produced the wonderful volume *Resilience Thinking* in 2006, which provides a more accessible read of the core concepts in *Panarchy*.

12. The term *panarchy* was first used by Belgian philosopher De Puydt in 1860. Sewell and Salter, 1995, applied a variation of the concept to global governance concepts. In the resilience literature, the term first appeared in Gunderson et al., 1995.

13. From this framework, panarchy makes five propositions about the way the world works and explores them in detail: (1) Biological and human entities form clumped structures of "anarchical" organization that create diversity while contributing to resilience and sustainability. (2) Sustainability is generated by interactions between nested sets of adaptive cycles arranged in a dynamic hierarchy in space and time. The theoretical foundations still rest on a focus of the development of alternative stable states at different levels of the hierarchy. (3) There are three kinds of change in a panarchy that facilitate different kinds of learning: incremental change and learning, abrupt change, and spasmodic change and transformational learning (see the more detailed discussion of the learning types in the previous chapter). (4) In studying complex systems, Holling's "rule of hand" suggests addressing only three to five variables. In understanding complex systems through the panarchy approach, less is more in eliminating noise and illuminating important dynamic interactions. (5) Self-organized ecological systems are structured by interactions between biota and physical variables within the arena of evolutionary change. Self-organization of human institutions builds social arenas for sustainable activities.

14. See figure 2.4 and associated discussion in chapter 2.

15. For a wonderfully rich and nuanced discussion of resilience and adaptive management, see Norton, 2005.

16. This approach as applied to the social sciences has been around since the 1920s, although the term itself first appeared in the work of MIT professor Kurt Lewis in the 1940s.

17. Based on conversations with former Everglades ecologist and adaptive management expert Steve Light, pers. comm., and a review of the early resilience literature.

18. Hilborn, pers. comm., via e-mail to the author in 2013. Also see Walters and Hilborn, 1978.

19. This book is affectionately known as "grey ghost" by practitioners due to its color. In Holling (1978) the order of editorship was chosen by lottery, with the organization of the book viewed as a collaborative effort. The full authorship included A. Bazykin, P. Bunnell, W. C. Clark, G. C. Gallopin, J. Gross, R. Hilborn, C. S. Holling, D. D. Jones,

R. M. Peterman, J. E. Rabinovich, J. H. Steele, and C. J. Walters. In 1986, Walter's volume *Adaptive Management of Renewable Resources* (known as the "green book" by practitioners) explored more deeply how adaptive management could be made operational.

20. As revealed by long-term studies by Moser and Moser (1986) that continued into the 1980s.

21. Although monitoring of photo points occurred in the Baker II example, and intensive experimentation and on-the-ground data collection on McKinney Flats, it is unfortunate that the degree of coordination was never attained to integrate the learning from these diverse experiences to the extent possible. This indicates that then, as now, social design still has overwhelming importance in maximizing the learning and benefits from these large-scale arenas of action.

22. The application of a social framework became much more widely applied later through Berkes and Folke's (1998) emphasis on socioecological systems that looked at the synergy of ecological and social sciences, and Francis Westley's (2008) focus on social innovation.

23. From Lee (1993), the phrase captures the essence of the adaptive management approach, which recognizes that one can choose to learn or not learn from experience. But because experience is expensive to acquire, practitioners should strive for maximum gain. The adaptive approach highlights the false dichotomy between research and practice. Without gaining the benefit of experience in a controlled and strategic way, mistakes are just repeated, and in our rapidly changing world there is no time to reinvent the wheel. The case studies in this book are replete with examples of mistakes being repeated because the system was not set up to profit from experience (see discussion of monitoring and funding in the borderlands as an example). Such approaches are really just single-loop learning, the pitfalls of which were discussed in chapter 4.

24. Current work explores watershed management in northern New Mexico in addition to continental-scale collaborative conservation programs with the Practitioner's Network for Large Landscape Conservation.

25. Holling et al. 1978; Walters 1986; Lee 1993; The origins of Gunderson et al. 1995. Attributed to conversation between Steve Light and Lance Gunderson in 1998.

26. I personally prefer *Barriers and Bridges* (1995) to the later volume *Panarchy* (2002) (and more recent overview books on resilience by Walker and Salt in 2006 and 2012), for although *Barriers and Bridges* is a more difficult read, and certainly less accessible to many readers, the underlying theory and assumptions are more clear, and many priceless nuggets of insight are missing from the later readings, which are simple, shorter, and more straightforward. Readers who are new to resilience science may wish to start with the two Walker and Salt books, then read Lee, 1993, and then *Panarchy*, before tackling *Barriers and Bridges,* which is a tough go if approached from scratch.

27. An outcome noted in published work by Jacobs and Westcoat, 2002, but also in conversations with USGS biologist David Mattson, who was part of the process.

28. See earlier discussion of resilience types by Handmer and Dovers, 1996, and discussion of learning loops in chapter 4.

29. The debate over mechanical trout removal versus changes in river conditions is illustrated in presentations by the USGS scientist David Mattson (unpublished).

30. Statement by Steve Light, pers. comm. Also see Light and Adamowski, 2012.

31. Lee, pers. comm.

32. The results were communicated in conversations with former Deputy Secretary of Interior Lynn Scarlett, who is familiar with the process. Also see Smith, 2011.

33. The science program from its beginnings was a balancing act between competing interests. Researchers were eventually caught in a tug of war between the MBG lead-

ership and the Diamond A, with the cutting of research access as much meant to send a message to the MBG leadership and The Nature Conservancy, as the researchers themselves. The actions were undertaken without consulting any of the partners who had more than a decade invested in the research.

34. Consider issues of power and core and periphery economies as illustrated in Steve Bunker's 1990 book *Underdeveloping the Amazon,* and in Durlauf and Blume, 2006. We explore it in the context of rangelands in Curtin et al., 2002, in which we look at the almost osmotic movement of wealth and how it drove overcapitalization of range in the 1880s and 1890s as European money fueled a speculative bubble that led to the great crash in cattle markets and ecosystems in the 1890s and again in the 1920s in the U.S. Southwest, permanently degrading much of the ecosystem.

35. The notion of power relationships, considered in the context of conservation and resource management, becomes even more fraught with complexity when applied across international boundaries, particularly those dividing wealthy, industrialized nations and the developing world. From a typical Western perspective, conservation is defined as protecting biodiversity, large landscapes, and warm fuzzy creatures in exotic locales. Yet there are fundamental issues of equity and rights that, from a developing world perspective, make conservation and resource management profoundly political. Historically, ecology has been the science of empire, a means developed by Western science to catalog the natural resources of other nations. Creating parks and reserves to protect and restrict access to endangered species has its roots in the European tradition of setting aside lands and game for the exclusive use of ruling elites. It was once illegal for an English commoner to shoot "the King's deer," even for sustenance in a time of need. In the developing world, parks and wildlife have remained a way for the elite to manipulate local peoples. In East Africa, colonial control and exclusivity of wildlife was so strong that native Kenyans still call wildlife the Swahili equivalent of "the queen's cattle," and parks such as Amboseli (discussed in chapter 1) are often referred to as "the queen's farm," illustrating the legacy of colonialism and power discontinuities still plaguing large-scale conservation. Also see Nadasky (2007, 2010) for discussions of power and the resilience paradigm.

36. See Clark et al., 1979, as an initial summary and Kates et al., 2001. This early work on ecological policy design is in many respects more sophisticated and nuanced than the later writing on resilience, which is more accessible to a broad audience. Some argue that to make resilience broadly accessible has meant simplifying it, and that in the process some of the elegance of the earlier work has been lost.

37. Michael, 1973, explores the shortcomings of planning processes and essentially makes this same case, as does Senge, 1990, both classic books on developing transitional approaches to thinking and problem solving.

38. http://thefreedictionary.com.

39. Curtin, 2014.

Chapter 6

1. See the discussion of ecologist Michael Rosenzweig and his 2003 book *Win-Win Ecology* in chapter 4.

2. Take, for example, William Newmark's work (1995, 1996) in parks in East Africa and North America, where even the largest reserves such as Yellowstone have experienced species decline through time.

3. See Olsson et al., 2006. The phrase is also widely used by practitioners across many areas of resource management.

4. As alluded to at the outset of the book, see Rittel and Webber, 1973, and Ison and Collins, 2008, for summaries of "wicked" problems. Also see Brown et al., 2010.

5. Homer-Dixon, 2006, nicely illustrates this through a discussion of the convergence of destructive synergies among population, climate change, and fossil fuels that currently face the planet.

6. Meadows, 1999, is a wonderfully succinct review of her ideas that she expanded on in her 2008 book *Thinking in Systems: A Primer.* Meadows's interest in levers arose out of noting that issues arise when planners try to push their way out of a problem by seeking simple, short-term solutions. "Leverage points are not intuitive," Meadows wrote. "Or if they are, we intuitively use them backward, systematically worsening whatever problems we are trying to solve." The reason for the counterintuitive outcome is that short-term benefits are rarely consistent with viable long-term solutions, and the timing of impacts in ecological systems rarely matches the pace of decision making, as in the example of society's inability to address climate change. Instead, impacts accumulate slowly, with sudden, dramatic environmental or social effects. This is illustrated by the adaptive cycle discussed in the previous chapter, in which the seemingly stable part of the cycle, the conservation phase, actually precedes collapse, just as the flow of a river is often smoothest just before an abrupt drop. As demonstrated in previous chapters, due to differences in initial conditions, making precise predictions regarding the dynamics of complex or wicked systems is all but impossible. Meadows was a founding member of the Club of Rome, the lead author of the landmark book *Limits to Growth,* and an early proponent of systems approaches to policy design. She died prematurely of a bacterial infection in 2001.

7. See Vaske and Whittaker, 2004.

8. Although goals should be attainable and quantifiable, they do not need to be explicit. So rather than seeking, for example, a specific outcome such as, say, an increase of 100 grams per square meter of usable forage following a prescribed fire (when one is really guessing at the correct number), one might instead seek an increase in vegetation diversity, an increase in the overall percentage of native grass cover, and a reduction in rate of erosion. Repeated experience may allow one to narrow in on more specific variables through time.

9. Senge, 1990.

10. See Brunner and Lynch, 2010, for a review of climate change governance. Great examples include the small cities of Burlington, Vermont, and Keene, New Hampshire, which have embraced climatic change in their planning processes. They even feature principles of climate change mitigation on the town websites.

11. As implied by Holling's adaptive cycle discussed in chapter 5, due to the intrinsic propensity of systems to change through time, it is much better to reduce pressure on systems when they are healthy and increasing than to try to reduce pressure during the down phase of the cycle. Yet this is essentially the exact opposite of how maximum sustained yield approaches tend to operate, all but dooming these management systems to decline and collapse, as the decline in the fishery discussed in chapter 4 illustrated.

12. See Curtin, 2014, for a more concise review of resilience design. Resilience design is essentially an emergent, complex systems approach to policy in which outcomes stem from relatively few ground rules.

13. See Norton, 2005. Norton, a philosopher by training, provides through numerous concise and well-developed arguments a series of concepts that are essentially proofs for why collaboration and adaptive management approaches are so necessary to building sustainability.

14. See Conklin's 2008 review of fragmentation in social systems.

15. The emergent properties of personal and group understanding demonstrate the power of collective thought and action and illustrate why these approaches are complex and time-consuming, but more sustainable than directed, command-control approaches to governance. As presented in previous chapters, from a thermodynamics and complex-systems perspective, the world is intrinsically dynamic, with emergent properties and dramatic thresholds. There is no real stability per se, only path dependency. Chaos theory demonstrates from a mathematical perspective that linear approaches are inherently problematic.

16. Bear in mind that in disasters or military campaigns where quick decisions are paramount, command-control approaches are essential. Again, it is a scale issue. The time of the decision-making process is related to the length of the solution. But even in more deliberative military campaigns that involve multiple players, there is a need for more integrated leadership such as Eisenhower displayed during World War II, or as has been noticeably absent in many recent military campaigns, such as in Afghanistan in the 2000s.

17. See Senge, 1990, and Leeuwis and Pyburn, 2002, for a discussion of the importance of social interactions in building sustainable institutions. This builds on experience in both organizational management and rural development.

18. An interesting attribute of sudden shifts in positive feedback loops is that they can trigger chaotic behavior. This happens when a system transitions faster than its negative loop can react. In pushing systems hard, exponential growth can be transitioned to oscillations, and then the smallest nudge can send it into chaos.

19. See McKinney and Johnson (2009) for an extensive review of governance in large systems.

20. Daniels and Walker, 2001, provided one of the most effective discussions within the context of natural resources by illustrating how to build collaborative processes. Also see Karl et al., 2007.

21. Choosing political expediency over effective process or ecological realities has been a recurrent problem undermining effective conservation in the rangelands and fisheries examples in this book, which illustrate why it is so important to explicitly consider ecological constraints and apply them to the policy process.

22. The importance of external leadership was especially striking in terms of the MBG's embracing experimental science. Very shortly after the departure of TNC leadership in coleading the MBG, scientists' access to Malpai meetings and other processes declined markedly. We went from being engaged in the very core of decision making with lengthy calls on a daily basis to becoming somewhat outsiders. Scientists would never again have strong, direct input into the decision-making process. Partly this was just a process of the local leadership taking over the reins, because full-time ranchers simply did not have time to devote to daily communications. At the same time there was a political element, in that science also represented something of a threat to local control, and the loss of external leadership represented a loss of cover for the science program, which began a slow atrophy of science-based approaches in the borderlands. Though it took a decade for the science program to end, the preconditions for loss were all set in the shift in the MBG power structure in the late 1990s, as it moved the researchers from insiders to outsiders of the decision-making system.

23. Donors often mark the success of collaborative processes by how rapidly the reins are turned over to the community, but this may, in fact, be exactly the wrong step. A structure combining local and nonlocal leadership combines the strengths of each and is more effective at sustaining local collaboratives. Collective impact's concept of a

backbone organization whose sole purpose is to sustain local organizations without an alternative political agenda, as discussed in chapter 3, may represent a critical step in building longevity by allowing organizations to transcend local politics.

24. See Homer-Dixon, 2006, which examines outcomes of peak oil and the need for dramatic changes in resource use.

25. See Checkland, 1985, as discussed in Ison and Collins, 2008.

26. Ecology is a curious hybrid—a soft science frequently masquerading as a hard one. The classic debates over theory of the 1960s to the present represent different assumptions about the nature of the system. Theorists such as MacArthur and colleagues took essentially a hard systems view and attempted a more reductionist approach; the systems thinkers of the era, such as Holling, viewed the world more as soft and complex. Resilience science is to a certain extent an effort to reconcile these different perspectives and link ecology with other disciplines.

27. Referred to in chapter 4 as Ashby's law. "Requisite variety" means that the greater the diversity of understanding, the greater the range of potential solutions.

28. See Berkes and Folke, 1998, and Berkes, 1999, for a description of local and community knowledge types and their significance.

29. A social perturbation, such as a change in institutional mandate, can have comparable positive or negative benefits—as do ecological processes.

30. Monitoring results must be documented and archived in a manner that is safe and accessible: If it is not accessible, it is not useful. But the data itself must be durable. For example, one of the best means of monitoring is through photography. Images that are now more than 100 years old have been of fundamental importance in understanding environmental change in arid and semi-arid ecosystems (Hastings and Turner, 1965; Turner et al., 2003). Modern digital photography is infinitely easier and cheaper to use than the old view cameras of a century ago, or even 35-mm film cameras of a few years ago. But archival methods for this digital media change almost yearly, making it uncertain whether this information will ever be around and accessible long enough to reach its intended purpose. For some of the most effective reviews of rangeland monitoring, see the work of Jeff Herrick and others at the Jornada Experimental Range in Las Cruces, New Mexico.

31. In the borderlands the McKinney Flat project was developed at the boundary of grassland and shrubland ecosystems. Near the center of ecosystems, it is difficult to see much response to perturbations, whereas near edges systems are much more dynamic (Curtin 2005, 2008).

32. An example from chapter 2 occurs in the grazing literature, where small-scale studies of grazing or fire yield fundamentally different outcomes. At small scales, these processes can be extremely damaging, as they disturb the entire area of interest. But at large scales, they are an intrinsic part of the system, adding richness and heterogeneity to the landscape. See figure 4.8 for another example.

33. Too often, separate committees or individuals consider different approaches to knowledge-gathering, creating fragmentation. A shortcoming of the Malpai science program has been precisely this. Monitoring and experimental research became segregated, and eventually each developed its own constituency; the lack of an integrated whole comes at the cost of the effectiveness of both.

34. A 2005 review of 425 of the most successful collaborative organizations associated with the White House Conference on Cooperative Conservation indicated that only a handful (including the MBG) followed these principles. See unpublished report by Karl Hess, 2005. http://govinfo.library.unt.edu/whccc.

35. Very often the opportunities to build social capital either by bringing people together in the research process or by providing credibility may outweigh the initial importance of the data itself. All of these considerations are also crucial components of design.

36. In essence, the "reorganization" phase of Holling's adaptive cycle. Mintzberg, 1987.

37. One of the clearest reviews of logic models is on a website developed by the Kellogg Foundation (http://www.wkkf.org/knowledge-center/resources/2006/02/WK -Kellogg-Foundation-Logic-Model-Development-Guide.aspx).

38. As discussed in chapter 2, in the 1980s, Quaker activist Jim Corbett helped refugees fleeing wars in El Salvador and Guatemala find their way into the United States. In the process he came to know borderlands ranchers and was one of the early outsiders to attend the Malpai meetings. The initial Malpai approach of openness and inclusiveness stems from Corbett's influence, and the group's principles include a strong Quaker influence.

39. Reviewed in more detail in Crocker, 2008.

40. Northwest Atlantic Marine Alliance. https://namanet.org/.

41. Ibid.

42. Yet, except for a few isolated references, finance is rarely considered. One of the few books to consider it at all in the context of developing effective process is Westley et al., 2006.

43. Conservation finance is the practice of raising and managing capital to support land, water, and resource conservation. Conservation financing options vary by source from public, private, and nonprofit funders; by type from loans to grants to tax incentives to market mechanisms; and by scale ranging from federal to state to local. See Clark, 2007, and the Conservation Finance Network at http://conservationfinancenetwork .org for more details.

44. This is a paradox that efforts such as collective impact are beginning to try to address, but it is unclear how widely adopted it will be, for funders will always need to be fluid and interests inevitably change through time.

45. Ironically, research and monitoring can exhibit the opposite pathologies. Although monitoring may initially seem a bargain, over the years as staff become more experienced and their salaries increase, the cost of sustaining programs can increase exponentially. In research the costs are front-loaded in setup, but through time work can become more streamlined, so costs can often go down.

46. The same issue occurred in fishery conservation on the coast of Maine, where here too it soon became clear that that was no chance of sustaining science-based programs long enough to answer the core questions needed to understand groundfish recovery. This would require at least a decade of effort, but there was unlikely to even be five years of funding.

47. The flow of resources appeared to go in one direction, with the lessons learned not readily communicated back to the funder, or if they were communicated (in the form of reports, etc.), they did not seem to feed into an adaptive process in which the funders learned from the outcomes of their efforts.

48. For example, a review of job websites such as Ecolog finds comparatively few multi-disciplinary jobs, but most grant competitions favor single-disciplinary proposals. Although NSF and other entities have greatly increased the number of transdisciplinary research programs, the breadth of the funding typically reduces actual funding opportunities for any individual researcher over more focused traditional, single-disciplinary approaches.

49. As discussed in earlier chapters, the use the phrase *socioecological systems* has increased

in recent years, but seems to result in a social science approach to ecology more than a true blending of perspectives, or when they are used, it is typically not well integrated. A point of this book is to show that coupling physical, natural, and social sciences can leverage all three, rather than resulting in a dumbing down of the different approaches to find a lower common denominator, as so often seems to be the case. Daniel Mc-Carthy's 2006 thesis is one of the most thoughtful efforts I have seen to blend these perspectives, which he does through the use of the phrase socio-ecological-epistemo-logical (SEE) systems. This makes clear that the systems are often not different, as much as the perception of them is. The point of this book is to clearly recognize that it is all one system, as framed by taking a large landscape (and seascape) approach.

50. See, for example, the work of the Cobscook Bay Resource Center in Maine discussed in chapter 3, which used a community kitchen to both fulfill their mission and sustain an income stream to support their work.

51. This perspective is not new, but reflects Aldo Leopold's comments in "A Land Ethic," which was the final chapter in *A Sand County Almanac*.

Literature Cited

Acheson, J. M. 1988. *The Lobster Gangs of Maine*. New England Quarterly, MIT Press, Cambridge, MA.

Agrawal, A. 2002. Common resources and institutional stability. pp. 41–86, in *The Drama of the Commons*, E. Ostrom, T. Dietz, N. Dolsak, P. C. Stern, S. Stonich, and E. U. Weber, eds. National Academy Press, Washington, DC.

Allen, T. F. H., and T. W. Hoekstra. 1992. *Toward a Unified Ecology*. Columbia University Press, NY.

Allen, T. F. H., and T. B. Starr. 1982. *Hierarchy: Perspectives for Ecological Complexity*. University of Chicago Press, Chicago.

Alverson, D. L., M. H. Freeberg, S. A. Murawski, and J. G. Pope. 1996. *A global assessment of fisheries bycatch and discards*. FAO Fisheries Technical Paper 339, United Nations, Rome.

Ames, T. 2004. Atlantic cod stock structure in the Gulf of Maine. *Fisheries* 29: 10–28.

Apollonio, S. 2002. *Hierarchical Perspectives on Marine Complexities*. Columbia University Press, NY.

Argyris, C., and D. Schön. 1974. *Theory in Practice. Increasing Professional Effectiveness*. Jossey-Bass, San Francisco.

Argyris, C., and D. Schön. 1978. *Organizational Learning: A Theory of Action Perspective*. Addison-Wesley, Reading, MA.

Armitage, D. 2008. Governance and the commons in a multi-level world. *International Journal of the Commons* 2(1): 7–32.

Ashby, W. R. 1962. Principles of the self-organized system. pp. 255–278, in *Principles of Self-Organization: Transactions of the University of Illinois Symposium*, H. Von Foerster and G. W. Zopf, Jr., eds., Pergamon Press, London, UK.

Augustine, D. J. 2010. Spatial versus temporal variation in precipitation in a semiarid ecosystem. *Landscape Ecology* 25: 913–925.

Baranger, M. 2011. Chaos-complexity-entropy: A talk for non-physicists. http://www .scribd.com/doc/157391647/Chaos-Complexity-and-Entropy-a-Physics-Talk-for-Non -Physicists-Michel-Baranger.

Bargh, J. A., and T. L. Chartrand, 1999. The unbearable automaticity of being. *American Psychologist* 54: 462–479.

Bateson, G. 1979. *Mind and Nature: A Necessary Unity*. E.P. Dutton, NY.

Beratan, K. K. 2007. A cognition-based view of decision processes in complex social-ecological systems. *Ecology and Society* 12(1).

Berkes, F. 1999. *Sacred Ecology: Traditional Ecological Knowledge and Management Systems*. Taylor & Francis, Philadelphia and London, UK.

Berkes, F., and C. Folke. 1998. *Linking Social and Ecological Systems*. Cambridge University Press, Cambridge, UK.

Berkes, F., T. P. Hughes, R. S. Steneck, J. A. Wilson, D. R. Bellwood, B. Crona, C. Folke, L. H. Gunderson, H. M. Leslie, J. Norberg, M. Nyström, P. Olsson, H. Österblom, M. Scheffer, and B. Worm. 2006. Globalization, roving bandits, and marine resources. *Science* 17: 1557–1558.

Boulding, K. E. 1950. *A Reconstruction of Economics*. J. Wiley, NY.

Bourque, B. J. 1995. *Diversity and Complexity in Prehistoric Maritime Societies: A Gulf of Maine Perspective*. Plenum Press, NY.

Brown, J. H. 1994. Organisms and species as complex adaptive systems: Linking the biology of populations with the physics of ecosystems. pp. 16–24, in *Linking Species and Ecosystems*. C. G. Jones and J. H. Lawton, eds. Chapman and Hall, London.

Brown, J. H. 1999. The legacy of Robert MacArthur: From geographic ecology to macroecology. *Journal of Mammalogy* 80: 333–344.

Brown, J. H., and A. Kodric-Brown. 1996. Biodiversity in the borderlands. *Natural History* 105: 58–61.

Brown, J. H., and W. McDonald. 1995. Livestock grazing and conservation on southwestern rangelands. *Conservation Biology* 8: 919–921.

Brown, J. H., D. W. Davidson, J. C. Munger, and R. S. Inouye. 1986. Experimental community ecology: The desert granivore system. pp. 41–62, in *Community Ecology*. J. Diamond and T. J. Case, eds. Harper & Row Publishers, NY.

Brown, J. H., T. J. Valone, and C. G. Curtin. 1997. Reorganization of an arid ecosystem in response to recent climate change. *Proceedings of the National Academy of Sciences* 94: 9729–9733.

Brown, V. A., J. A. Harris, and J. Y. Russell. 2010. *Tackling Wicked Problems: Through the Transdisciplinary Imagination*. Earthscan Press, NY.

Brunner, R. D., and A. H. Lynch. 2010. *Adaptive Governance and Climate Change*. American Meteorological Society, University of Chicago Press, Chicago.

Bunker, S. G. 1990. *Underdeveloping the Amazon: Extraction, Unequal Exchange, and the Failure of the Modern State*. University of Chicago Press, Chicago.

Cable, D. R. 1965. Damage to mesquite, Lehmann love grass, and black grama by a hot June fire. *Journal of Range Management* 18: 326–329.

Cahill, L., and McGaugh, J. L. 1998. Mechanisms of emotional arousal and lasting declarative memory. *Trends in Neuroscience* 21: 294–299.

Cain, S. A. 1938. The species-area curve. *American Midland Naturalist* 19: 573–581.

Campbell, D. T. 1969. Reforms as experiments. *American Psychologist* 24: 409–429.

Canby, W. C. 1981. *American Indian Law in a Nutshell*. West, St. Paul, MN.

Capra, F. 1996. *The Web of Life: A New Synthesis of Mind and Matter*. Harper Collins Publishers, NY.

Carpenter, S. 1996. Microcosm experiments have limited relevance for community and ecosystem ecology. *Ecology* 77: 677–680.

Checkland, P. 1985. From optimizing to learning: A development of systems thinking for the 1990s. *The Journal of the Operational Research Society* 36: 757–767.

Child, B., and M. Lyman. 2005. *Natural Resources as Community Assets: Lessons from Two Continents*. Sand County Foundation, Madison, WI, and Aspen Inst., Aspen, CO.

Choi, J. S., K. T. Frank, B. D. Petrie, and W. C. Leggett. 2005. Integrated assessment of a large marine ecosystem: A case study of the devolution of the eastern Scotian Shelf, Canada. *Annual Review of Oceanography and Marine Biology* 43: 47–67.

Clark, S. 2007. *A Field Guide to Conservation Finance*. Island Press, Washington, DC.

Clark, W. C. 2002. Adaptive management, heal thyself. *Environment* 44: 2.

Clark, W., C. S. Holling, and D. D. Jones. 1975. *Towards a Structural View of Resilience*. Working Paper International Institute for Applied Systems Analysis. WP-75-96.

Clark, W. C., D. D. Jones, and C. S. Holling. 1979. *Lessons for Ecological Policy Design: A Case Study of Ecosystem Management*. R-10-B. Institute of Resource Ecology, University of British Columbia, Vancouver, BC.

Clements, F. E. 1916. *Plant Succession: An Analysis of the Development of Vegetation*. Carnegie Institution of Washington.

Conklin, J. 2008. Wicked problems and social complexity. http://cognexus.org/wpf/wick edproblems.pdf.

Corson, T. 2004. *The Secret Life of Lobsters*. Harper Collins, NY.

Crocker, M. 2008. *Sharing the Ocean: Stories of Science, Politics, and Ownership*. Tilbury House, Gardner, ME.

Curtin, C. G. 2002a. Science and community-based conservation in the Malpai borderlands. *Conservation Biology* 16: 880–886.

Curtin, C. G. 2002b. Cattle grazing, rest, and restoration in arid lands. *Conservation Biology* 16: 840–842.

Curtin, C. G. 2005. Linking complexity, conservation, and culture in the Mexico/US borderlands. pp. 235–258, in *Natural Resources as Community Assets: Lessons from Two Continents*. B. Child and M. West Lyman, eds. Sand County Foundation, Madison, WI, and Aspen Institute, Aspen, CO.

Curtin, C. G. 2008. Emergent outcomes of the interplay of climate, fire, and grazing in a desert grassland. *Desert Plants,* December. Tucson.

Curtin, C. G. 2010. The ecology of place and natural resource management: Lessons from marine and terrestrial ecosystems. pp. 251–274, in *The Ecology of Place: Contributions of Place-Based Research to Ecological Understanding*. I. Billick and M. Price, eds. Columbia University Press, NY.

Curtin, C. G. 2014. Resilience design: Toward a synthesis of cognition, learning, and collaboration for adaptive problem solving in conservation and natural resource stewardship. *Ecology and Society* 19(2): 15. http://dx.doi.org/10.5751/ES-06247-190215.

Curtin, C. G., and J. H. Brown. 2001. Climate and herbivory in structuring the vegetation of the Malpai borderlands. pp. 84–94, in *Vegetation and Flora of La Frontera: Vegetation Change along the United States-Mexico Boundary*. C. J. Bahre and G. L. Webster, eds. University of New Mexico Press, Albuquerque.

Curtin, C. G., and S. Hammitt. 2012. Outcomes of social-ecological experiments: Cognitive interpretation of the impact of changes in fishing gear type on ecosystem form and function. pp. 457–474, in *Restoring and Sustaining Lands: Coordinating Science, Politics, and Action*. H. A. Karl, M. Flaxman, J. C. Vargas-Moreno, and P. Lynn Scarlet, eds. Springer-Verlag, The Netherlands.

Curtin, C. G., and J. P. Parker. 2014. Foundations of resilience science. *Conservation Biology* 4: 912–923.

Curtin, C. G., and D. Western. 2008. Rangelands as a global conservation resource: Lessons from cross-cultural exchange between pastoral cultures in East Africa and North America. *Conservation Biology* 22: 870–877.

Curtin, C. G., T. C. Frey, D. A. Kelt, and J. H. Brown. 1999. On the role of small mammals in mediating climatically driven vegetation change. *Ecology Letters* 3: 309–317.

Curtin, C. G., N. Sayre, and B. Lane. 2002. Transformation of the Chihuahua borderlands: Biodiversity conservation and landscape fragmentation in desert grasslands. *Environmental Science and Technology* 218: 55–68.

Cyert, J., and J. Marsh. 1963. *A Behavioral Theory of the Firm*. Prentice-Hall, NYC.

Daniels, J., and G. Walker. 2001. *Working through Environmental Conflict*. Praeger, Santa Barbara, CA.

Darwin, C. 1859. *The Origin of Species*. John Murray, London, UK.

De Puydt, P. E. 1860. *Panarchy*. Revue Trimestreille, Bruxelles.

Durlauf, S., and L. Blume. 2006. *The Economy as an Evolving Complex System III*. Oxford University Press, NY.

Easterby-Smith, M., L. Araujo, and J. G. Burgoyne. 1999. *Organizational Learning and the Learning Organization: Developments in Theory And Practice*. SAGE Publications, London.

Eddington, A. 1930. *The Nature of the Physical World.* Macmillan, NY.

Fel'dbaum, A. A. 1961. Dual control theory, Parts I and II. *Automation and Remote Control* 21(9): 874–880 and 21(11): 1033–1039. (Russian originals dated September 1960, pp. 1240–1249, and November 1960, pp. 1453–1464.)

Flood, R. L., and N. R. A. Romm. 1996. *Diversity Management: Triple Loop Learning.* John Wiley and Sons, NY.

Frank, K., B. Petrie, J. Choi, and W. Leggett. 2005. Trophic cascades in a formerly cod-dominated ecosystem. *Science* 308: 1621–1623.

Frank, K. T., B. Petrie, and N. Shackell. 2007. The ups and downs of trophic control in continental shelf ecosystems. *Trends in Ecology and Evolution* 22: 236–242.

Funtowicz, S. O., and J. R. Ravetz. 1993. Science for the post-normal age. *Futures* 25: 739–755.

Gell-Mann, M. 1994. *The Quark and the Jaguar: Adventures in the Simple and the Complex.* W. H. Freeman and Company, NY.

Golley, F. B. 1993. *A History of the Ecosystem Concept in Ecology: More than the Summary of the Parts.* Yale University Press, New Haven.

Gottfried, G. J., L. G. Eskew, C. G. Curtin, and C. B. Edminster. 1999. *Toward integrated research, land management, and ecosystem protection in the Malpai Borderlands.* Conference summary, 6–8 January 1999, Douglas, AZ. Proceedings, Rocky Mountain Research Station, p. 10. Fort Collins, CO.

Graham, J., S. Engle, and M. Recchia. 2002. *Local knowledge and local stocks.* A report of the Centre for Community-Based Management. Extension Department, St. Francis Xavier University, Antigonish, Nova Scotia, Canada.

Gunderson, L. H., and C. S. Holling. 2002. *Panarchy: Understanding Transformations in Human and Natural Systems.* Island Press, Washington, DC.

Gunderson, L. H., C. S. Holling, and S. S. Light. 1995. *Barriers and Bridges to the Renewal of Ecosystems and Institutions.* Columbia University Press, NY.

Gunderson, L. H., C. R. Allen, and C. S. Holling. 2010. *Foundation of Ecological Resilience.* Island Press, Washington, DC.

Handmer, J. W., and S. R. Dovers. 1996. A typology of resilience: Rethinking institutions for sustainable development. *Organization Environment* (December 9): 482–511.

Hanleybrown, F., J. Kania, and M. Kramer. 2012. Channeling change: Making collective impact work. *Stanford Social Innovation Review* 2012: 1–8.

Hardin, G. 1968. The tragedy of the commons. *Science* 162: 1243–1248.

Harrington, J. M., R. A. Myers, and A. A. Rosenberg. 2005. Wasted fishery resources: Discarded by-catch in the USA. *Fish and Fisheries* 6: 350–361.

Hastings, J. R., and R. M. Turner. 1965. *The Changing Mile: An Ecological Study of Vegetation Change with Time in the Lower Mile of an Arid and Semi-Arid Region.* University of Arizona Press, Tucson.

Hatfield, H., T. Ecoffey, and J. Stone. 2013. *A Guide to Managing Tribal Buffalo Herds.* Intertribal Buffalo Council, Rapid City, SD.

Hess, K., 2005. White House Conference on Collaborative Conservation. http://govinfo .library.unt.edu/whccc/agenda.html.

Hess, K., Jr., and J. L. Holechek. 1995. Policy roots of land degradation in the arid region of the United States: An overview. *Journal of Environmental Monitoring and Assessment* 37: 10–19.

Hilborn, R. C. 2004. Sea gulls, butterflies, and grasshoppers: A brief history of the butterfly effect in nonlinear dynamics. *American Journal of Physics* 72: 425–427.

Hilliard, G. 1996. *A Hundred Years of Horse Tracks.* High Lonesome Books, Silver City, NM.

Hobbs, R. J., S. Arico, J. Aronson, J. S. Barson, P. Bridgewater, V. A. Cramer, P. R. Epstein, J. J. Ewel, C. A. Klink, A. E. Lugo, D. Noron, D. Ojima, R. M. Richardson, E. W. Sanderson, F. Valladares, M. Vila, R. Zamora, and M. Zobel. 2006. Novel ecosystems: Theoretical and management aspects of the new ecological world order. *Global Ecology and Biogeography* 15: 1–7.

Homer-Dixon, T. 2006. *The Upside of Down: Catastrophe, Creativity, and the Renewal of Civilization.* Island Press, Washington, DC.

Holland, J. 1995. *Hidden Order: How Adaptation Builds Complexity.* Helix Books, NY.

Holling, C. S. 1959. The components of predation as revealed by a study of small-mammal predation of the European pine sawfly. *The Canadian Entomologist* 91: 293–320.

Holling, C. S. 1973. Resilience and stability of ecological systems. *Annual Review of Ecology and Systematics* 4: 1–23.

Holling, C. S., ed. 1978. *Adaptive Environmental Assessment and Management.* John Wiley & Sons, Chichester, UK.

Holling, C. S. 1986. The resilience of terrestrial ecosystems: Local surprise and global change. pp. 292–317, in *Sustainable Development of the Biosphere.* W. C. Clark and R. E. Munn, eds. Cambridge University Press, Cambridge, UK.

Holling, C. S. 2006. A journey of discovery. rs.resalliance.org/author/buzz-holling/.

Holling, C. S., and A. D. Chambers. 1973. Resource science: The nurture of the infant. *BioScience* 23: 13–20.

Holling, C. S., and G. K. Meffe. 1996. Command and control and the pathology of natural-resource management. *Conservation Biology* 10: 328–337.

Hughes, T. P. 1994. Catastrophes, phase-shifts, and large-scale degradation of a Caribbean coral reef. *Science* 265: 1547–1551.

Hutchins, E. 1995. *Cognition in the Wild.* MIT Press, Cambridge, MA.

Innis, H. 1977. *The Fur Trade in Canada: An Introduction to Canadian Economic History.* University of Toronto Press, Toronto.

Ison, R., and K. Collins. 2008. Public policy that does the right thing, rather than the wrong thing righter. In *Analysing Collaborative and Deliberative Forms of Governance.* Crawford School of Economics and Government, Australian National University, Canberra.

Jackson, J. B., M. X. Kirby, W. H. Berger, K. A. Bjorndal, L. W. Botsford, B. J. Bourque, R. H. Bradbury, R. Cooke, J. Erlandson, J. A. Estes, T. P. Hughes, S. Kidwell, C. B. Lange, H. S. Lenihan, J. M. Pandolfi, C. H. Peterson, R. S. Steneck, M. J. Tegner, and R. R. Warner. 2001. Historical overfishing and the recent collapse of coastal ecosystems. *Science* 293: 629–638.

Jacobs, J. W., and J. L. Westcoat, Jr. 2002. Managing river resources: Lessons from Glen Canyon Dam. *Environment* 44: 8–19.

Jantasch, E. 1980. *The Self-Organizing Universe: Scientific and Human Implications of the Emerging Paradigm of Evolution.* Pergamon Press, NY.

Jones, A. 2000. Effects of cattle grazing on North American arid ecosystems: A qualitative review. *Western North American Naturalist* 60: 155–164.

Jorgensen, S. E., and B. D. Fath. 2004. Application of thermodynamic principles in ecology. *Ecological Complexity* 4: 267–280.

Kania, J., and M. Kramer. 2011. Collective impact. *Stanford Social Innovation Review.* Winter.

Karl, H., L. E. Susskind, and K. H. Wallace. 2007. A dialogue, not a diatribe. *Environment* 49: 20–34.

Kates, R., et al. 2001. Sustainability science. *Science* 292: 641–642.

Kaufman, S. A. 2008. *Reinventing the Sacred: A New View of Science, Reason, and Religion.* Basic Books, NY.

Kauffman, M., J. F. Brodie, and E. S. Jules. 2010. Are wolves saving Yellowstone's aspen? A landscape-level test of a behaviorally mediated trophic cascade. *Ecology* 91: 2742–2755.

Kearney, J. 2005. Community-based fisheries management in the Bay of Fundy: Sustaining communities through resistance and hope. pp. 83–100, in *Natural Resources as Community Assets: Lessons from Two Continents*. Child, B. and Lyman, M.W. eds. Aspen Institute, Washington, DC.

Keen, M., V. A. Brown, and R. Dyball. 2005. *Social Learning in Environmental Management: Toward a Sustainable Future*. Earthscan, London, UK.

Kurlansky, M. 1997. *Cod: A Biography of the Fish That Changed the World*. Penguin Books, NY.

Lee, K. N. 1993. *Compass and Gyroscope*. Island Press, Washington, DC.

Leeuwis, C., and R. Pyburn. 2002. *Wheelbarrows Full of Frogs: Social Learning in Rural Resource Management*. Koninklijke van Gorcum, The Netherlands.

Leopold, A. 1933. *Game Management*. Charles Scribner & Sons, NY.

Leopold, A. 1949. *A Sand County Almanac*. Oxford, NY.

Levin, S. A. 1999. *Fragile Dominion: Complexity and the Commons*. Perseus Books, Reading, MA.

Lewontin, R.C. 1969. The meaning of stability. *Brookhaven Symposia on Biology* 22: 13–24.

Light, S. S., and J. Adamowski. 2012. *Restoring Lands-Coordinating Science, Politics, and Action*. H. A. Karl, L. Scarlett, J. C. Vargas-Moreno, M. Flaxman, eds. Springer, NY.

Lindeman, R. L. 1942. The trophic-dynamic aspect of ecology. *Ecology* 23: 399–418.

List, R., G. Ceballos, C. G. Curtin, G. P. Gogan, J. Pacheco, and J. Truett. 2007. Conservation and ecology of a Chihuahuan bison herd: Implications for conserving a species and their habitat. *Conservation Biology* 21: 1487–1494.

Lorenz, E. 1963. Deterministic non-periodic flow. *Journal of Atmospheric Sciences* 20: 130–141.

Loverde, L. 2005. Learning organizations and quadruple loops of feedback. *Ingenierias, Enero-Marzo* 8: 29–36.

Ludwig, D. 2001. The era of management is over. *Ecosystems* 4: 758–764.

MacArthur, R. 1955. Fluctuations of animal populations and a measure of community stability. *Ecology* 36: 533–536.

MacArthur, R. H. 1972. *Geographical Ecology: Patterns in the Distribution of Species*. Princeton University Press, Princeton, NJ.

MacArthur, R. H., and E. O. Wilson. 1967. *The Theory of Island Biogeography*. Princeton University Press, Princeton, NJ.

Maestas, J. D., R. L. Knight, and W. C. Gilgert. 2002. Cows, condos, or neither. What's best for rangeland ecosystems? *Rangelands* 24: 36–42.

Manchester, W. 1992. *A World Lit by Fire: The Medieval Mind and Renaissance*. Little, Brown & Company, NY.

Mandelbrot, B. 1967. How long is the coast of Britain? Statistical self-similarity and fractional dimension. *Science* 156: 636–638.

March, J. G., and H. A. Simon. 1958. *Organizations*. John Wiley and Sons, NY.

Margalef, R. 1968. *Perspectives in Ecological Theory*. University of Chicago Press, Chicago.

Masten, A., K. Best, and N. Garmezy. Resilience and development: Contributions from the study of children who overcome adversity. 1990. *Development and Psychopathology* 2: 425–444.

Maturana, H. R., and F. J. Varela, 1987. *The Tree of Knowledge: The Biological Roots of Human Understanding*. Shambala, Boston, MA.

May, R. M. 1972. Will a large complete system be stable? *Nature* 238: 413–414.

McAllister, R. R. J., I. J. Gordon, M. A. Janssen, and N. Abel. 2006. Pastoralists' responses to variation of rangeland resources in time and space. *Ecological Applications* 16: 572–583.

McCarthy, D. D. P. 2006. *A Critical Systems Approach to Socio Ecological Systems: Implications for Social Learning and Governance.* Thesis, Doctor of Philosophy in Planning. University of Waterloo, Waterloo, Ontario, Canada.

McClaran, M. P., and T. R. Van Devender. 1995. *The Desert Grassland.* University of Arizona Press, Tucson.

McKinney, M. J., and S. Johnson. 2009. *Working Across Boundaries: People, Nature, and Regions.* The Lincoln Institute, Cambridge, MA.

Meadows, D. H. 1999. *Leverage Points: Places to Intervene in a System.* The Sustainability Institute, Norwich, VT.

Meadows, D. H. 2008. *Thinking in Systems: A Primer.* Sustainability Institute, Norwich, VT.

Michael, D. 1973. *Learning to Plan and Planning to Learn.* Jossey-Bass, NY.

Milchunas, D. 2006. *Response of plant communities to grazing in the southwestern United States.* USDA Forest Service, Rocky Mountain Research Station, General Technical Report 169.

Milchunas, D. G., and W. K. Lauenroth. 1993. Quantitative effects of grazing on vegetation and soils over a global range of environments. *Ecological Monographs* 63: 327–366.

Mintzberg, H. 1987. Crafting strategy. *Harvard Business Review* 65: 66–75.

Morgan, G. 1988. *Riding the Waves of Change.* Jossey Bass, San Francisco, CA.

Moser, P., and W. Moser. 1986. Reflections on the MAB-6 Obergurgl project and tourism in an alpine environment. *Mountain Research and Development* 6: 101–118

Muchiru, A. N., D. J. Western, and R. S. Reid. 2008. The role of abandoned pastoral settlements in the dynamics of African large herbivore communities. *Journal of Arid Environments* 72: 940–952.

Muldavin, E. H., D. I. Moore, S. L. Collins, K. R. Wetherill, and D. C. Lightfoot. 2008. Above-ground net primary production dynamics in a northern Chihuahuan Desert ecosystem. *Oecologia* 155: 123–132.

Mumford, M. D. 2002. Social innovation: Ten cases from Benjamin Franklin. *Creativity Research Journal* 14: 253–266.

Mwangi, E., and E. Ostrom. 2009. Top-down solutions: Looking up from East Africa's rangelands. *Environment* Jan–Feb: 34–44.

Myers, R. A., and B. Worm. 2003. Rapid worldwide depletion of predatory fish communities. *Nature* 423: 280–283.

Nadasdy, P. 2007. Adaptive co-management and the gospel of resilience. pp. 208–227, in *Adaptive Co-Management: Collaboration, Learning, and Multilevel Governance.* D. Armitage, F. Berkes, N. Doubleday, eds. University of British Columbia Press, Vancouver.

Nadasdy, P. 2010. Resilience and truth: Response to Berkes. *MAST* 9: 41–45.

National Research Council. 1994. *Rangeland health: New methods to classify, inventory, and monitor rangelands.* Committee on Rangeland Classification, Board of Agriculture, National Research Council, National Academies Press, Washington, DC.

Nature, editorial. 2011. Fix the PhD. *Nature* 472: 259–260.

National Research Council. 2008. *Progress toward restoring the Everglades: The second biennial review, 2008.* The National Academies Press, Washington, DC.

Newmark, W. D. 1995. Extinction of mammal populations in western North American national parks. *Conservation Biology* 9: 512–526.

Newmark, W. D. 1996. Insularization of Tanzanian parks and the local extinction of large mammals. *Conservation Biology* 10: 1549–1556.

Norberg, J., and G. S. Cumming. 2008. *Complexity Theory for a Sustainable Future.* Columbia University Press, NY.

Norton, B. G. 2005. *Sustainability: A Philosophy of Adaptive Ecosystem Management.* University of Chicago Press, Chicago.

Odum, H. T. 1983. *Systems Ecology: An Introduction.* John Wiley, NY.

Odum, E. P., and H. T. Odum. 1953. *Fundamentals of Ecology.* Saunders, Philadelphia, PA.

O'Leary, M. W. 1996. *Maine Sea Fisheries: The Rise and Fall of a Native Industry: 1830–1890.* Northeastern University Press, Chicago.

Olsson, P., L. H. Gunderson, S. R. Carpenter, P. Ryan, L. Lebel, C. Folke, and C. S. Holling. 2006. Shooting the rapids: Navigating transitions to adaptive governance of social-ecological systems. *Ecology and Society* 11: 18.

O'Neill, R. V., D. L. DeAngelis, J. B. Waide, and T. F. H. Allen. 1986. *A Hierarchical Concept of Ecosystems.* Princeton University Press, Princeton, NJ.

Ostrom, E. 1990. *Governing the Commons: The Evolution of Institutions for Collective Action.* Cambridge University Press, Cambridge, UK.

Ostrom, E. 2007. A diagnostic approach for going beyond panaceas. *Proceedings of the National Academy of Sciences* 104: 15181–15187.

Pauly, D., V. Christensen, J. Dalsgaard, R. Froese, and F. Torres. 1998. Fishing down marine food webs. *Science* 279: 860–863.

Pound, R., and F. E. Clemens. 1898. A method of determining the abundance of secondary species. *Minnesota Botanical Studies* 2: 19–24.

Preston, F.W. 1962. The canonical distribution of commonness and rarity: Part I. *Ecology* 43 185–215 and 410–432.

Prigogine, I. 1996. *The End of Certainty.* Free Press, NY.

Prigogine, I., and Stengers, I. 1984. *Order Out of Chaos: Man's New Dialogue with Nature.* Bantam Books, NY.

Proulx, J. 2008. Some differences between Maturana and Varela's theory of cognition and constructivism. *Complicity* 5: 11–26.

Quammen, D. 1996. *The Song of the Dodo: Island Biogeography in an Age of Extinction.* Touchstone, NY.

Ravetz, J. 2002. The post-normal science precautionary principle. http://www.nusap.net/.

Remley, D. 2000. *Bell Ranch: Cattle Ranching in the Southwest, 1824–1947.* Yucca Tree Press, Las Cruces, NM.

Ripple, W. J., and R. L. Beschta. 2011. Trophic cascades in Yellowstone: The first 15 years after wolf reintroduction. *Biological Conservation* 145: 205–213.

Rittel, H. W. J., and M. M. Webber. 1973. Dilemmas in a general theory of planning. *Policy Sciences* 4: 155–169.

Rosenberg, A. A., W. J. Bolster, K. E. Alexander, W. B. Leavenworth, A. B. Cooper, and M. G. McKenzie. 2005. The history of ocean resources: Modeling cod biomass using historical records. *Frontiers of Ecology and the Environment* 3: 84–90.

Rosenzweig, M. L. 1995. *Species Diversity in Space and Time.* Cambridge University Press, Cambridge, UK.

Rosenzweig, M. L. 2003. *Win-Win Ecology: How the Earth's Species Can Survive in the Midst of Human Enterprise.* Oxford University Press, Oxford, UK.

Russell, 2009. *Report on the impact of the 2009 drought on the South Rift Valley ecosystem and comparison to the Amboseli ecosystem.* Report to African Conservation Centre and South Rift Association of Land Owners.

Sala, O. E., L. A. Gherardi, L. Reichmann, E. Jobbágy, and D. L. Peters. 2012. Legacies of precipitation fluctuations on primary production: Theory and data. *Philosophical Transactions of the Royal Society of London. Series B: Biological Sciences* 367: 313–314.

Sayre, N. 2005. *Working Wilderness: The Malpai Borderlands Group and the Future of the Western Range.* Rio Nuevo Publishers, Tucson.

Scheffer, M., S. Carpenter, and B. de Young. 2005. Cascading effects of overfishing marine systems. *Trends in Ecology and Evolution* 20: 579–581.

Schindler, D. W. 1998. Replication versus realism: The need for ecosystem-scale experiments. *Ecosystems* 1: 323–334.

Schneider, E. D., and J. J. Kay. 1994. Life as a manifestation of the second law of thermodynamics. *Mathematical and Computer Modeling* 19: 25–48.

Schumpeter, J. A. 1942. *Capitalism, Socialism and Democracy*. Harper, NY.

Sears, P. B. 1935. *Deserts on the March*. University of Oklahoma Press, Norman, OK.

Senge, P. 1990. *The Fifth Discipline: The Art and Practice of the Learning Organization*. Doubleday Currency, NY.

Sewell, J. P., and M. B. Salter. 1995. Panarchy and other norms for global governance: Boutros-Ghali, Rosenau, and beyond. *Global Governance* 3: 373–382.

Simon, H. 1947. *Administrative Behavior: A Study of Decision-Making Processes in Administrative Organization*. Macmillan, NY.

Simon, H. 1962. The architecture of complexity. *Proceedings of the American Philosophical Society* 106: 467–482.

Sissenwine, M. P. 1984. The uncertain environment of fishery scientists and managers. *Marine Resource Economics* 1: 1–30.

Smith, C. B. 2011. Adaptive management on the central Platte River—science, engineering, and decision analysis to assist in the recovery of four species. *Journal of Environmental Management* 92: 1414–1419.

Smith, E. 1990. Chaos in fisheries management. *Maritime Anthropological Studies* 3: 1–13.

Soulé, M. E., and B. Wilcox. 1980. *Conservation Biology: An Evolutionary-Ecological Perspective*. Sinauer Associates, Sunderland, MA.

Steneck R. S., and J. A. Wilson. 2010. A fisheries play in an ecosystem theater: Challenges of managing ecological and social drivers of marine fisheries at nested spatial scales. *Bulletin of Marine Science* 86: 387–411.

Steneck, R. S., J. Vavrinec, and A. V. Leland. 2004. Accelerating trophic-level dysfunction in kelp forest ecosystems of the western Gulf of Maine. *Ecosystems* 7: 323–332.

Steneck, R. S., T. P. Hughes, J. E. Cinner, W. N. Adger, S. N. Arnold, F. Berkes, S. A. Boudreau, K. Brown, C. Folke, L. Gunderson, P. Olsson, M. Scheffer, E. Stephenson, B. Walker, J. Wilson, and B. Worm. 2011. Creation of a gilded trap by the high economic value of the Maine lobster fishery. *Conservation Biology* 25: 904–912.

Swetnam, T. W., and J. L. Betancourt. 1998. Mesoscale disturbance and ecological response to decadal climatic variability in the American Southwest. *Journal of Climate* 11: 3128–3147.

Tainter, J. 1990. *The Collapse of Complex Societies*. Cambridge University Press, Cambridge, UK.

Tansley, A. G. 1935. The use and abuse of vegetational terms and concepts. *Ecology* 16: 284–307.

Taylor, P. J. 2005. *Unruly Complexity*. University of Chicago Press, Chicago.

Tilt, W., C. Conley, M. James, J. C. Lynn, T. A. Muñoz-Erickson, and P. Warren. 2008. Creating successful collaborations in the West: Lessons from the field. pp. 1–23, in *The Colorado Plateau III: Integrating Research and Resources Management for Effective Conservation*. C. Van Riper III and M. Sogee, eds. University of Arizona Press, Tucson.

Tobey, R. 1981. *Saving the Prairies: The Life Cycle of the Founding School of American Plant Ecology, 1895–1955*. University of California Press, Berkeley.

Truett, J.C., M. Phillips, K. Kunkel, and R. Miller. 2001. Managing bison to restore biodiversity. *Great Plains Research: A Journal of Natural and Social Sciences* 11: 123–144

Turner, R. M., R. H. Webb, J. E. Bowers, and J. R. Hastings. 2003. *The Changing Mile Revisited*. University of Arizona Press, Tucson.

Vaske, J., and D. Whittaker, 2004. Normative approaches to natural resources. pp. 283–294, in *Society and Natural Resources: A Summary of Knowledge*. M. J. Manfredo, J. J. Vaske, B. L. Bruyere, D. R. Field, and P. J. Brown, eds. Modern Litho, Jefferson, MO.

Von Bertalamffy, K. L. 1968. *General System Theory: Foundations, Development, Applications.* George Braziller, NY.

Walker, B., and D. Salt. 2006. *Resilience Thinking.* Island Press, Washington, DC.

Walker, B., and D. Salt. 2012. *Resilience Practice.* Island Press, Washington, DC.

Walters, C. J. 1986. *Adaptive Management of Renewable Resources.* Macmillan, NY.

Walters, C. J. 1998. Evaluation of quota management studies for developing fisheries. *Canadian Journal of Fisheries and Aquatic Sciences* 55: 2691–2705.

Walters, C. J., and R. Hilborn 1975. *Adaptive control of fishing systems.* International Institute for Applied Systems Analysis, WP-75-114.

Walters, C. J., and R. Hilborn. 1978. Ecological optimization and adaptive management. *Annual Review of Ecology and Systematics* 9: 157–188.

Wang, C. L., and P. K. Ahmed, 2001. *Creative quality and value innovation: A platform for competitive success. Integrated Management—Conference Proceedings of the 6th International Conference of ISO9000 and TQM*, Scotland, April, pp. 323–329.

Weaver, W. 1948. Science and complexity. *American Scientist* 36: 536–547.

Wessels, T. 2006. *The Myth of Progress: Toward a Sustainable Future.* University of Vermont Press, Burlington, VT.

Western, D. 1997. *In the Dust of Kilimanjaro.* Island Press/Shearwater Books, Washington, DC.

Western, D. 2000. Conservation in a human-dominated world. *Issues in Science and Technology,* Spring: 53–61.

Western, D., R. Groom, and J. Worden. 2009. The impact of subdivision and sedentarization of pastoral lands on wildlife in an African savanna ecosystem. *Biological Conservation* 142: 2538–2546.

Westley, F. 1995. Governing design: The management of social systems and ecosystem management. pp. 391–427, in *Barriers and Bridges to the Renewal of Ecosystems and Institutions.* Gunderson L.H., Holling C.D., Light S., eds. Columbia University Press, NY.

Westley, F. 2008. The social innovation dynamic. http://tinyurl.com/l89dwpq.

Westley, F., B. Zimmerman, and M. Q. Patton. 2006. *Getting to Maybe: How the World Is Changed.* Random House, Toronto.

Wilson, E. O., and G. E. Hutchinson. 1989. *Robert Helmer MacArthur, April 7, 1930 – November 1, 1972.* Biographical Memoirs, v. 5, National Academy Press, Washington, DC.

Wilson, J. A. 2002. Scientific uncertainty, complex systems, and the design of common-pool institutions. pp. 327–360, in *The Drama of the Commons.* E. Ostrom, T. Dietz, N. Dolsak, P. C. Stern, S. Stonich, and E. U. Weber, eds. National Academy Press, Washington, DC.

Wilson, J. A. 2006. Matching social and ecological systems in complex ocean fisheries. *Ecology and Society* 11(1): 9. http://www.ecologyandsociety.org/vol11/iss1/art9/.

Wilson, J., L. Yan, and C. Wilson. 2007. The precursors of governance in the Maine lobster fishery. *Proceedings of the National Academy of Sciences* 104: 15212–15217.

Wolf, T. 2001. *The Malpai Borderlands Group: Science, community, and collaborative management.* Workshop on Collaborative Resource Management in the Interior West. Red Lodge Clearing House, Boulder, CO.

Woodward, C. 2005. *The Lobster Coast: Rebels, Rusticators, and the Struggle for a Forgotten Frontier.* Viking Press, NY.

Worster, D. 1994. *Nature's Economy,* 2nd ed. Cambridge University Press, Cambridge, UK.

Wright, S. 1932. The roles of mutation, inbreeding, crossbreeding, and selection in evolution. *Proceedings of the Sixth International Congress on Genetics.* pp. 355–366.

About the Author

Charles G. Curtin is a senior fellow at the Center for Natural Resources and Environmental Policy on the University of Montana campus in Missoula, and a landscape ecologist at the Center for Large Landscape Conservation in Bozeman, MT. He works at the nexus of science and policy, with a long-term interest in environmental change, large-scale socioecological experiments, and conservation design, focusing primarily on community-based conservation, large-scale experimental science, and restoration of rangeland ecosystems. He helped design some of the largest place-based collaborative research programs on the continent, including the million-acre Malpai Borderlands conservation area and cross-site studies spanning the Intermountain West. He has also worked with fisheries policy and comanagement through development of the 750,000-square-mile Downeast Initiative in the western Atlantic and anadromous fish restorations on the coast of Maine. He has helped established academic programs in governance and policy design at the Massachusetts Institute of Technology (MIT) and Antioch University with a focus on collaborative approaches to climate change adaptation and mitigation. Curtin has also worked internationally, coordinating large-landscape collaborative conservation projects in East Africa and the Middle East.

Index

Figures/photos/illustrations are indicated by "f."

Island Press | Board of Directors